PROMISING THE EARTH

'This book is much more than a fascinating story about Friends of the Earth. It's about the evolution of the green movement in Britain and the new environmental agenda – an agenda that is driving society forward more effectively than any other. It's a must-read for all those interested in campaigning for change.'

Charles Secrett, *Executive Director, Friends of the Earth*

'FoE is indeed an all-embracing organisation, a microcosm of the whole environment movement in the UK. Its history both casts light on and reflects 25 years of environmental ideas, campaigns and projects'.

Jonathon Porritt

'Think globally – act locally.' True to its slogan, Friends of the Earth is recognised for its achievements in every area of environmental activity, from parish pump to Parliament.

Published to mark the 25th anniversary of Friends of the Earth, *Promising the Earth* presents a vivid account of the experiences, achievements and expectations of Britain's first environmental pressure group. It reveals, for the first time, the inside story of pioneering environmental campaigns – from the first anti-whaling and fur trade campaigns of the early 1970s to the anti-nuclear demonstrations at Windscale and Sellafield and the most recent anti-road protests at Twyford Down and Newbury.

At a time when environmental groups worldwide are taking stock after the Rio Earth Summit, *Promising the Earth* offers important lessons for green campaigners around the world as they face the challenges of the new millennium.

Robert Lamb is an independent writer and researcher who has headed the information services of both the World Conservation Union and the United Nations Environment Programme.

PROMISING THE EARTH

Robert Lamb
In collaboration
with Friends of the Earth

London and New York

First published 1996
by Routledge
11 New Fetter Lane, London EC4P 4EE

Simultaneously published in the USA and Canada
by Routledge
29 West 35th Street, New York, NY 10001

Typeset in Sabon by
Florencetype Ltd, Stoodleigh, Devon

Printed and bound in Great Britain by
Biddles Ltd, Guildford and King's Lynn

British Library Cataloguing in Publication Data
A catalogue record for this book is available from the British Library

Library of Congress Cataloging in Publication Data
Lamb, Robert 1949–
Promising the earth/Robert Lamb; in collaboration with Friends of the Earth.
p. cm.
Includes bibliographical references and index.
1. Friends of the Earth – History. 2. Green movement – Great Britain – History. 3. Environmental protection – Great Britain – History.
4. Sustainable development – Great Britain – History.
I. Friends of the Earth. II. Title.
GE199.G7L36 1996
363.7′0576′0941 – dc20 96–7556
CIP

ISBN 0–415–14443–4 (hbk)
ISBN 0–415–14444–2 (pbk)

Contents

Figures

Foreword

Jonathon Porritt

When I applied to become Director of Friends of the Earth in 1984, I proceeded with some caution. I'd heard so many good and not so good things about it. One close colleague advised against with these memorable words, 'It's not just another job in another organisation; it's a cause, and it takes people over, body, mind and soul'. He was right. It did, and it was probably the best decision of my life.

FoE is indeed an all-embracing organisation, a microcosm of the whole environment movement in the UK. Its history both casts light on and reflects twenty-five years of environmental ideas, campaigns and projects. No other environmental organisation represents quite such a broad church in terms of the range of support it draws on, from the heart of the Establishment to the very fringes of alternative society. None has covered a wider range of issues, including blazing the trail on a host of popular campaigns such as whaling and endangered species which others have subsequently been able to capitalise on.

No other organisation operates at so many different levels: through its individual supporters; through a network of 250 local groups; through its campaigns and a separate charitable trust at the national level; and through a network of national FoE groups that is still growing. What's more, I would doubt that any other organisation lays claim to quite so many philosophical and political antecedents, and probably only Greenpeace uses such a wide range of tactics to get its message across.

Being such a broad church has given FoE many strengths over the years, especially in the occasional but inevitable lean periods. But it brings its own problems, precipitating ferocious, if mercifully infrequent, bouts of in-fighting over both strategy and structure. Every incoming Director of FoE longs to find ways of managing such internal ructions out of the system. But it comes with the territory, and the hassle it causes is a price that's worth paying for what is perhaps FoE's greatest strength: the passionate commitment of its staff and volunteers in both the national organisation and its local groups.

In one unequal tussle after another (unequal in terms of financial resources, that is) it is this commitment that has given FoE the edge over backsliding government departments, misbehaving businesses and fence-sitting quangos. It is this commitment which has often enabled it to turn failure into symbolic success. Schweppes, for instance, is even now resisting the introduction of reusable bottles, but FoE's historic 'bottle dump' on its doorstep still retains its power as an image of more efficient energy use and resource management. And it's also what enables FoE to pick itself up after any failures, find a way of turning even those disappointments into object lessons for the future. There's no clearer example of this than the inquiry into the THORP reprocessing plant at Windscale in the late 1970s. FoE won the arguments hands down – as every subsequent decision over the next twenty years has amply demonstrated – but the decision went against it. It was a crushing blow. But almost unnoticed by those coping with the aftermath, it encouraged the development of one crucial aspect of FoE's campaigning that has never really been properly recognised.

These days, 'solutions-oriented' campaigning is all the rage. Now that the case against our wasteful, unsustainable ways has been embraced by literally all and sundry, the emphasis has shifted to promoting the policies, processes and technologies to get us out of the mess. FoE has been doing precisely that (especially on the energy and transport fronts) for the best part of fifteen years – even though for much of that time the media couldn't have cared less about such an approach.

On the whole, however, FoE cannot complain of its treatment at the hands of the media. In the 1970s, when nobody was listening very much, it managed to find a number of 'multipliers' who forced FoE's issues into the public domain. In the 1980s, the media counted on FoE to fill a quite astonishing political vacuum arising out of the utter failure of the opposition parties in the UK to run with the green agenda. And in the 1990s, FoE's strategic repositioning on the very much broader post-Rio agenda of sustainable development has greatly strengthened its intellectual credibility – without yet bearing fruit in terms of column inches!

This relatively benign relationship with the media may of course have something to do with the fact that former FoE employees would seem to be out and about in every conceivable neck of the green woods – in local and national government, academia and education, business and the voluntary sector. It has provided a formidable training ground over the years, giving people skills and a depth of understanding that have proved utterly invaluable to the movement as a whole.

But what I come back to time after time in reading this book is the creative tension between FoE's role as dispassionate purveyor of hard-edged empirical evidence about the state of the Earth, and its role as a crusading advocate of a vision which intuitively ascribes value to the Earth and its creatures in ways that cannot be constrained through simplistic cost-benefit analysis. Getting the science right has always been a watchword of FoE's effectiveness, but there has also always been a realm of meaning and civic purpose for FoE beyond that which may be deemed economically viable or scientifically irrefutable. And there is little doubt that the general mood is swinging back in favour of just such a values-driven approach.

So is FoE truly an organisation for all seasons? Will some even longer in the tooth ex-Director be penning a Foreword in 2021 to mark the fiftieth anniversary? Given its protean capacity to reinvent itself, and in the process to help the whole environment movement do the same, I suspect that's more than likely. And I can't help thinking that the world will be a better place because of it.

Preface

This book's appearance is scheduled to coincide with the twenty-fifth anniversary in the UK of Friends of the Earth. Its scope is not, however, limited to 'official biography' nor to the British domestic scene. It sets out to take a fresh look at the global surge of environmental concern and consciousness since 1960, a tale of our times that no single organisation's record can be expected to paraphrase. It will, however, locate the inside story of a selection of FoE's campaigns and experiences within this broader canvas. These insights bear intimate and eloquent witness to a process still underway – the maturing of the grassroots pro-environment movement as a positive force for change and care in the natural world, in politics, law, technology and commerce *and* in the ideas and lives of people and communities everywhere.

I warmly thank the current and former FoE campaigners and managers who spared time and trouble to place their knowledge, ideas and reminiscences on tape for the book. They all spoke candidly and speculated freely about their experiences and expectations. Also useful were notes prepared by Tom Burke, Czech Conroy and Chris Church to mark earlier anniversaries. No restriction was placed on my assignment; nothing was bowdlerised and nobody muzzled. All I was asked to do was tell the story in an alive and frank yet measured and fair way, as an informed onlooker with no axe to grind, or at least nothing too heavyweight in the axe line.

I value this enlightened brief and shall do my best to live up to it as well, I hope, as turning out an entertaining and relaxed narrative – my own prescription for the work. Books on the history of movements do tend to make eyes glaze over and if this is no exception it was not for want of trying.

I ought to make it clear at once that I hold no brief for the 'contrarian' case against pro-environment groups and their philosophies in the mass. The contrarian stance involves weaving a home-made straw figure of the green movement as if its definition were fixed in time and custom, then gleefully taking this unrecognisable scarecrow apart. I hope this book will manage to avoid such glib and self-important posturing.

On the other hand I do not cherish any illusion that green organisations lead charmed lives free from wrong turns, rash misjudgements or colliding personalities. Anyone who has worked for a corporate (or any) institution knows such flaws are endemic to them all. It would be dim to expect ethical groups to make a better job of dodging them than the rest, especially if they happen to inhabit so volatile and vexed a tideline as society's interactivity with nature.

I doubt, even so, whether there are many institutions with more good-humoured determination to learn by experience or with hearts more definitively in the right place than Friends of the Earth and its peer groups in the independent green movement around the world. I take my hat off to them all.

Chronology

1971 FoE launched in the UK – and dumps 1,500 non-returnable bottles outside Schweppes' HQ.

1972 FoE wins ban on imports of leopard, cheetah and tiger skins.

1973 FoE's campaigning forces UK ban on all baleen whale products. RTZ abandons its mining plans for Snowdonia.

1974 FoE's lobbying wins higher insulation standards under Health and Safety law. The Government cancels its plans to build 32 new PWRs after FoE's campaign highlights cost, safety and environmental risks.

1975 FoE Durham is the first group to win Government funding for practical energy conservation schemes.

1976 The Endangered Species Act 1976, first drafted by FoE, becomes law.

1977 FoE persuades the Department of Energy to set up the first national domestic insulation scheme. The Conservation of Wild Creatures and Wild Plants Act 1977, first drafted by FoE and RSNC, becomes law.

1978 Otter hunting is banned in England and Wales. FoE's amendments on cycling to the Transport Bill are accepted. An FoE Nottingham campaign persuades Nottingham Council to reject plans for a Woolco out-of-town hypermarket and Woolco to drop major nationwide hypermarket expansion plans. The leather industries agree to stop using sperm whale oil.

1979 FoE prosecutes House of Sears for selling leopard skins. The Ministry of Defence announces it will no longer use whale products, and British Rail reverses a ban on bikes. The IWC is persuaded to designate the Indian Ocean a whale sanctuary.

1980 FoE reveals secret Government negotiations to expand the nuclear programme. The Government agrees to halt a sell-off of allotments.

1981 Otter hunting is banned in Scotland. The Government establishes Marine Nature Reserves, thanks to WWF and FoE lobbying.

1982 FoE's consumer campaign wins a Europe-wide ban on all whale products. The IWC agrees a 10-year moratorium on commercial whaling, following a joint FoE, Greenpeace and WWF campaign.

1983 FoE launches an eventually successful campaign to ban straw-burning. The Labour and Liberal parties, along with the TGWU, pass an FoE-initiated resolution calling for greater wildlife and countryside protection.

1984 Peaceful direct action campaigns with local people save wildlife sites at Lippenhoe Marshes on the Broads, Uddens Heath SSSI in Bournemouth, and Primrose Hill near Biggin Hill. Wheatley FoE saves Otmoor, inspiration for *Alice Through the Looking Glass*.

1985 The London Dumping Convention agrees to an indefinite ban on all radioactive waste dumping at sea, a success for FoE and Greenpeace. Ely FoE saves fields adjoining Ely Cathedral.

1986 Local people working with FoE stop proposed nuclear waste dumps at Billingham, Bradwell, Fulbeck and Killingholme. The Government protects Halvergate Marshes in Norfolk, after an FoE and CPRE campaign. RSPB and FoE win protection for Duich Moss SSSI. The pesticides cyhexatin and dinoseb are banned.

1987 An FoE campaign stops Coca-Cola Foods from destroying tropical forests in Belize for citrus plantations. FoE's boycott campaign persuades McDonald's and Kentucky Fried Chicken to stop using CFC-blown cartons.

1988 FoE's boycott persuades UK aerosol manufacturers to phase out CFCs. FoE's Radiation Monitoring Unit exposes hundreds of radioactive sites around the country. The Government withdraws permits for nitrate pollution of drinking water supplies.

1989 The pesticides dieldrin and aldrin are banned. FoE reveals Government proposals to privatise nature reserves; the plans are rapidly abandoned. The Government withdraws nuclear stations from its privatisation plans and cancels three new PWRs.

1990 Scott Paper abandons plans to raze 300,000 hectares of Indonesian tropical forest for tissue paper plantations, following FoE's campaign. Governments agree to phase out ozone depleter 1,1,1-trichloroethane. The UK government abandons its proposed £3.5 billion roadbuilding programme for London, following FoE and local community campaigns.

1991 FoE takes the Government to the High Court over breaches of European drinking water law; one month before the hearing, the Government withdraws its case and speeds up the removal of pesticides from drinking water. FoE calls on the Government to triple its target for renewable energy production; the Government subsequently doubles support for renewable energy projects and increases supply targets.

1992 FoE wins a partial ban on the pesticides atrazine and simazine. Parliament passes the Traffic Calming Act 1992, drafted by FoE. DIY chain B&Q leads four other major DIY chains in agreeing to halt mahogany sales, following FoE's campaign. The European Court finds the UK government guilty of failing to comply with standards for nitrates in drinking water.

1993 FoE's Radiation Monitoring Unit discovers 33 radiation 'hot spots' along the coast between Bangor and Sellafield. Nuclear Electric closes Trawsfynydd nuclear station. The major DIY chains agree to stop selling peat from threatened wildlife sites, following a campaign by the Peatlands Campaign Consortium, of which FoE is a member. The campaign by PARC and FoE to save Oxleas Wood SSSI succeeds.

1994 FoE International wins funding for a radio network for Amazonian tribes threatened by loggers. The Body Shop and FoE supply geographical positioning equipment to demarcate tribal land boundaries. The UK government drops 49 new road schemes, many actively opposed by FoE for years. FoE breaks the story that leads to the exposure of a major British aid scandal in Pergau, Malaysia.

1995 The Government announces the end for nuclear power subsidies – and that no more nuclear stations will be built – a notable victory for FoE and Greenpeace. The Home Energy Conservation Act, campaigned for by the Green Party, the Association for the Conservation of Energy and FoE, is passed. After five years' campaigning, mahogany imports have fallen by over 50 per cent. The Government abandons another 77 road schemes, many opposed by FoE and local groups for years.

1

Another brick in the wall

These are times that try men's souls. Tyranny, like hell, is not easily conquered yet we have this consolation with us, that the harder the conflict, the more glorious the triumph. What we obtain too cheap, we esteem too lightly. Let it be told to the future that in the depth of winter, when nothing but hope and virtue could survive, the city and the country, alarmed at one common danger, came forth to meet and repulse it.

– Tom Paine, *The American Crisis*, 1776

The fight for Twyford Down in Hampshire was one of the most bitter battles over roadbuilding schemes in Britain in the 1980s and after. Its outcome has been gouged into the English landscape.

Twyford Down today is a white gash in the grey-green chalk uplands some 20 miles east of Stonehenge. A trick of the language calls this hill country downland; its gentle yet definite lines suggest living shapes. But here the motorway cutting severs and brutally truncates the whaleback form of the Down. Traffic whizzes through, on its way to or from London to the south coast. Who could look at Twyford Down now and doubt that the battle to save it was lost?

Yet Twyford has come to be regarded by many as the defeat that won the war. It saw the most radical and active protests ever mounted in Britain in defence of natural places. And it cradled a virtually new breed of protester.

Established groups such as Friends of the Earth had plenty to say about schemes like the completion of the M3 motorway through Twyford Down. Not surprisingly, as the downland in the M3's path formed the western extremity of the South Downs Area of Outstanding Beauty, included part of the St Catherine's Hill Site of Special Scientific Interest (SSSI) and sheltered two scheduled ancient monuments. At the foot of the down, where the motorway would emerge from the 400-foot-wide cutting on a 30-foot-high embankment, lay the River Itchen water-meadows, another SSSI. Together these provided valuable habitat and landscape features for uncommon species of flowering plants, birds and butterflies, an island of natural diversity surrounded by a sea of monocrop agriculture.

For environmentalists, however, the idea of cleaving a six-lane highway through this natural and cultural heritage raised deeper, more universal issues. Some had argued for almost three decades against the 'petrochemical culture' that had permeated postwar life in the industrialised countries of the global North and was fast spreading to the South. Skyrocketing economic growth had collided with self-evident natural limits. Unless alternatives arose, the backlash from such 'progress' would degrade human life and maybe blight all life on Earth.

In many areas of debate, the Establishment paid at least lip-service to the pro-environment case. In Britain, even Margaret Thatcher joined a general chorus of 'we are all environmentalists now', despite her regular insistence that profitability and economic growth were the only credible yardsticks of progress in human development. Yet in Thatcher's Britain, an economic boomtime was giving rise, even as she spoke, to a phenomenal increase in car ownership.

By 1992 there were more than 20 million cars in Britain, compared with 2 million back in 1952; around a third of this growth took place in the 1980s. To serve this 'Great Car Culture' as Thatcher termed it, a massive programme of new roadbuilding and road enlarging had been undertaken by the Department of Transport. The M3 extension was just part of it.

The issue was not new. FoE had campaigned persistently for environmentally responsible transport alternatives almost since it set up shop in Britain. But anti-road protest had recently taken on renewed meaning as a front line of pro-environment activity and controversy.

As well as the disappearance of the countryside and wildlife habitats, the roads juggernaut inevitably towed more major issues in its wake, like energy and atmospheric pollution. For Friends of the Earth, it offered a chance to bring these global issues into the open and to realise the organisation's unique potential for locally-based activism backed up by a national, centralised knowledge base and publicity machine. As an anti-road arena, Twyford shone.

The scheme was the latest in a long line of official assaults on SSSIs and other supposedly protected areas. In 1990 alone, 324 British SSSIs had been damaged or downsized as a result of development projects, mainly new roads and private housing projects, and the tally was climbing year by year. The regulations and financial levers that were meant to ensure their protection had never been consistently effective, despite (or maybe because of) the room for local compromise they were designed to leave. But until Twyford it was still almost unheard-of for an SSSI to be wiped off the face of the earth at one stroke.

For many people living in the nearby city of Winchester, the scheme threatened to demolish a favourite local beauty spot and prized recreational amenity. By no stretch of the imagination could the local protesters be dismissed as leftist rabble-rousers. On the whole they were prosperous and solidly middle-class townspeople, so dominated by inveterate Tory voters that it was sometimes said in jest that the Conservative vote in Winchester had to be weighed, not counted.

Their concern over the fate of Twyford Down was tied to traditional values, but local opposition to the M3 cutting through the Down could not be dismissed as mere NIMBYism. Locals had been campaigning against different proposed routes for the M3 for twenty years.

The local campaign group that manifested itself in the final battle was the Twyford Down Association (TDA), constituted formally in 1991 but effectively in operation since 1985. Its founding members could be stereotyped as traditional upholders of 'Middle England' values, but their fight led them through both the UK and European courts, saw them reject the values and policies of the Conservative government that many of them had voted in, and took its toll both personally and financially. The overriding theme of the Twyford Down campaign was to be this radicalising of all involved.

An appraisal published by the Department of Transport several years previously had attempted to prove that time saved on millions of quicker car journeys was a reasonable trade-off against the disfigurement of the local landscape. The same figures were used to rubbish a compromise solution that had some support among all the groups that opposed the scheme, which was to put the road *under* the Down, through a tunnel. Many locals lost patience when this alternative was dismissed. The area's prosperity owed much to people's pride in a quality landscape. Didn't that make 'economic sense' of the tunnel compromise?

The Twyford Down Association used all the conventional channels open to them: public inquiries, political connections, press coverage and the courts. When their High Court challenge finally failed, and with legal action from the EC only a distant possibility, they invited FoE into the campaign to attempt to raise public and political heat before the diggers actually began to tear up the downland.

As local campaigner Barbara Bryant wrote in her recent book *Twyford Down, Roads, Campaigning and Environmental Law*,

> It was quite clear that time was running out for Twyford Down. Although the intervention of the European Commission remained a possibility it seemed increasingly less likely, and we had nearly exhausted much of our energy, and most of our resources. Andrew Lees (Friends of the Earth's Campaigns Director), and Friends of the Earth, brought us campaigning expertise and resources when they were most needed.

Lees and former FoE Director Jonathon Porritt had been helping the TDA with its strategy and legal actions for some months previously, but in early 1992 FoE became visibly active on the ground. In February FoE staff and local campaigners occupied a bridge that was due to be demolished as part of the preliminary clearance works on the way to Twyford

Down itself. Robin Maynard, now FoE 's Local Campaigns Director but at the time the Countryside Campaigner, explains,

> The railway bridge occupation was a symbolic declaration of our intent to take action. When the police told us to move on the next day or face arrest we did so. Nothing sacred was immediately under threat, but we wanted to put the Government on notice that when it came to the SSSIs, FoE intended to stand up and be counted.

The Twyford Down Association also invited a second group to join in – and they occupied a second bridge just up the track from FoE's. This group was Earth First! 'Group' is not quite the right word, for unlike FoE, with its formal structures, offices, staff, local groups and long lists of supporters, Earth First! represents a philosophical banner under which individuals can gather to take part in direct actions on behalf of the environment. Inspired by the writings of nature-centred 'Deep Ecologists' such as Arne Naess and George Sessions, and more specifically by Edward Abbey's 1975 novel *The Monkey Wrench Gang*, Earth First! and its direct action tactics had first emerged in the USA. Twyford Down marked the debut of this organisation (or 'disorganisation', as it prefers to be called) in Britain.

However, throughout the month of February 1992, the battlefield of Twyford Down was primarily defended by the more traditional tactics of FoE and the Twyford Down Association. At dawn on the last day of February, the first day when work could legally begin on the SSSI sites, FoE set up a camp and a 'chain of protection' on the River Itchen watermeadows. The chain strung across the path of the motorway was inspired by the words of Sir William Wilkinson, outgoing Chair of the Nature Conservancy Council, as he attacked the Conservative Government for splitting the conservation body into three and thereby weakening the cause of nature conservation.

Wilkinson looked to British citizens to defend the countryside that the Government no longer cared for:

> To achieve success, nature conservation has to win the hearts and the interest of people throughout our land . . . Local enthusiasm will need to be harnessed to an effort of national will. It has to be carried out throughout the UK . . . wherever there is a community to enjoy our country's magnificent heritage. It is a protective chain which we must forge around our land; if any link in it fails, the chain itself will snap.

FoE's chain did not hold back the bulldozers for long, but it did kick-start events and provide a focus for the more radical tactics and robuster 'D-locks' (hardened steel bicycle locks used by protesters for attaching themselves to machinery) that came later.

Figure 1 A child adds her piece to the 'chain of protection' protest, Twyford Down © Times Newspapers

NO ROAD
SAVE OUR SITES
Friends of the Earth
There must be a better way!
F.O.E say DON'T LET TWFORD DOWN

For a fortnight in cold, damp February weather, FoE held the water-meadows and kept contractors' diggers off the Down itself with a second chain. But on 13 March 1992, an injunction was served against the organisation. Lawyers advised that to break it would risk contempt of court proceedings, which could result in fines of up to a quarter of a million pounds. This would bankrupt the organisation. FoE left the site, leaving a breach that was soon filled by the direct action protesters.

Whilst injuncted off the actual path of the motorway, FoE sustained a campaigning presence. When the bulldozers moved onto the water-meadows, three days after police had bolt-cut FoE and TDA protesters from the 'chain of protection', local people gathered in protest. Robin Maynard recalls,

> Friends of the Earth had the force of the law used against us when we were seeking to uphold both UK and European environmental law. It was incredibly frustrating not be able to continue the protest, but we were determined to show the Government that they couldn't stop local people's opposition. The rally on the watermeadows – which produced the fantastic sight of local people surrounding the bulldozers and stopping work – was organised anonymously by FoE with the TDA. The images of 'Middle England' in revolt, with an ex-Battle of Britain Squadron Leader and families standing up against the Government, were exactly those we wished to communicate. Unfortunately we didn't communicate the role we were attempting to play properly to the media or to those in the direct action movement. It still pains me that FoE missed the trick of understanding how to work collaboratively and complementarily at Twyford, but then we weren't alone in that.

In strategic terms, the campaign to save Twyford Down seemed doomed as the European Commission dropped its case against the UK Government. Then a tactical voting initiative, led by Jonathon Porritt, failed to split the Conservative vote at the general election and the Conservative Government was returned with what it saw as a renewed mandate to build roads. The new, less coldly strategic, more emotional and spontaneous style of campaigning by Earth First! and others challenged the rationality of the traditional campaigning approach. They generated enormous media interest and imposed a huge financial burden upon building a road whose route had been chosen as the cheapest option – running through countryside was deemed to be of little economic value.

The direct actions gained momentum in the summer, when topsoil was removed from the downland leaving a white chalk scar, visible for miles. This evidence of the injury being inflicted on 'Mother Earth' galvanised opposition and anger, as people saw for the first time how destructive the motorway would be.

The tactics used at Twyford went way beyond the non-violent 'media theatre' type of protest made famous by Greenpeace, FoE and other radical green groups active since the 1970s. Along with Earth First!, the protesters on the Down included the newly-fledged British pressure group Road Alert and squads of freelancers, including a few day-tripping hunt saboteurs. Then there were families from New Age Traveller convoys and groups of Crusties, raggedy and dreadlocked, part of a teenage underclass that appeared in small-town Britain in the wake of the Thatcher era.

These drifters had nothing, and therefore nothing to lose, yet they were mobile and in the vigour of youth. Many of them were also literate, educated and confident that, despite their apparent lack of power, they could come and go freely in the land and the land was there for them as a common property resource. They mostly got by on begging, scavenging and fortnightly Social Security handouts. But it was a point of honour to many Crusties and Travellers that they had some kind of vitalising bond with the land and owed it something in return.

The Dongas Tribe was the most visible of various small-scale subcultures that turned up at Twyford to live this idea out. The Dongas, who took their name from the mediaeval trackways on the Down, identify themselves in earnest with pre-scientific cultures like the Aboriginal or Native Australian societies of Oceania, the rainforest Amerindians of the Amazon Basin or the Native American people of Canada and the USA. At Twyford they turned out to be the wildest of all the informal subgroups. They also took some of the most demanding risks.

The professional campaigners at FoE greeted the arrival of Earth First! and its tatterdemalion legions with mixed feelings, though it raised the media stakes considerably. Roger Higman, FoE's Transport Campaigner, recalls

> They seemed incredibly young. Clearly they cared but we thought they were very confused, very inexperienced. We'd been used to dealing with local residents, all basically ordinary people in the area. And our own activists were mostly thirty-somethings, people with jobs that also let them pursue campaigns on subjects they cared about. All of a sudden you had kids, scruffy kids. They had hardly any knowledge about anything else in the roads programme, this was just the one road they'd heard of. They wanted to fight for every tree.

A stretch of imagination is required to picture the genial, bespectacled thirty-something Higman as a scruffy kid himself. But in the early 1980s he had more than just spiked hair in common with the confused-but-caring defenders of Twyford Down. He was then a student living in Wanstead, East London, in a row of houses owned by the Department of Transport and let temporarily to housing collectives. Their ultimate fate was demolition to make way for an urban motorway scheme, one of many schemes proposed for London and then undergoing feasibility studies.

Higman grew up in Oxford, home of one of Rover's biggest UK factories. 'I was angry about the car from a very early age', he relates. 'I was a keen cyclist from the age of five. I used to get cut up by cars when I was cycling to school practically every day.' His bike also carried him into the woodlands and wetlands of the Oxfordshire countryside, where he developed a lifelong passion for bird-watching.

Higman first heard of FoE in his Wanstead days, when he read about a campaign the group had mounted in his home patch. The Department of Transport had plans to build the M40 motorway across Otmoor in Oxfordshire, originally a lowland marsh and still, though heavily drained, a wild-looking place that attracts many birds which nest or feed in wet grasslands. Otmoor had inspired parts of Lewis Carroll's classic story *Through the Looking Glass*.

Wheatley FoE's coordinator, Joe Weston, had the brilliant idea of buying a meadow on Otmoor, smack on the road's intended route, and dividing it into one-metre squares. The group sold titles to 1,000 of these plots to people all over the world. The Department of Transport was unable to purchase the land, and the meadow was saved in 1985. However the loophole in the law which Weston had spotted and successfully exploited was then closed. The next time his trick was tried the Department of Transport overrode it.

Higman was impressed. 'It was so clever. And it really struck me then that what was happening to the street I lived in in London was the same thing that was threatening the countryside. I felt: thank God somebody's opposing this.'

After graduating, Higman volunteered his services at FoE's London headquarters. There followed what he calls the 'first peak' of anti-road campaigning in Britain's recent history, centred on London and sparked off by the publication in summer 1987 of a set of official 'Assessment Studies' outlining a hugely ambitious roadbuilding plan for the increasingly traffic-clogged capital.

Along with other groups and working within the coalition known as ALARM (All London Against the Road Menace), FoE became heavily involved in large-scale protest demonstrations and actions all over London, with great success. The Department of Transport abandoned nearly all their road schemes for London in 1989, dumping the rest in March 1990.

For FoE, which had started out on its transport campaigning in London with a string of local schemes for installing bike routes and 'traffic calming' road humps, the strength of feeling stirred up by the Assessment Studies came as a surprise. Higman recalls:

> All of a sudden we went from sitting in meetings of four or five people talking about bus routes, or twenty people talking about traffic calming, to a situation where we had thousands getting very angry about roadbuilding. I suppose what happened then is that a whole generation of campaigners got turned onto the idea that roads were where the real power lay. If we got involved with that we thought we'd change transport policy for ever.

But in June 1989, while these campaigners were focused on London, the Department of Transport suddenly doubled the roads programme in the rest of the country.

In 1990 Peter Witt, Director of the British Road Federation, predicted, 'The environmentalists are not going to enjoy the same success across the rest of the country as they've enjoyed in London.' He reckoned without FoE's ability to organise and coordinate local protests, nationwide. Higman says,

> I'd realised from what happened in London that as soon as the roads get stopped there's a problem of what to do next. Suddenly people who were out on the streets are sitting at home watching telly. So I figured there were great opportunities in fighting the national roads programme. I applied for the job of Transport Campaigner at Friends of the Earth and got it.

FoE gave their new campaigner the brief to do what had been done in London all over the country. The campaign had a string of minor successes over local road schemes south of Birmingham. They set a fine lead. FoE Birmingham, for long one of the most effective of the local groups, appointed their own campaigner, who built alliances with neighbourhood groups. The road schemes perished.

But out in the country things seemed to be going, if anywhere, more the Department of Transport's way. Other schemes were added to the roads programme. Studies for an outer London orbital motorway, outside the M25, were in the programme. FoE groups on the east coast united to fight a proposed East Coast motorway. Others joined with the residents' campaign group PARC to demonstrate at Oxleas, an ancient woodland on a proposed East London River Crossing route, later to be dropped after many acrimonious battles.

But the breadth of FoE's anti-roads campaigning won them few friends at Twyford. 'The direct action activists at Twyford didn't understand that we were fighting other roads too, which were just as important.' Higman recalls. 'They tended to call up and say: "Well? Why aren't you down here? They're coming in tomorrow!"'

'Because of the injunction, we had to back out early from the NVDA process at Twyford,' says Higman ruefully, 'and we got a shafting for it. But at the same time we were opening another front entirely, at the M25. From our point of view it was just as important.'

The proposed doubling in width of the M25, London's orbital motorway, to fourteen lanes each way, was meeting fierce opposition in the normally quiescent southern Home Counties, reckoned the most prosperous heartland of the Tory-voting middle class, a leafy habitat of monied professionals and retired dignitaries. It was a situation not unlike Twyford, but in this case the combination of resourceful and affluent local supporters and the ample lead time marked it out as a more likely winner. FoE switched much of its attention and resources to the M25 controversy, ultimately playing a major part in helping local people force the scheme's abandonment.

For FoE, the Twyford Down affair and the M25 were two battles in the same war. But many of those who stayed on at Twyford didn't see it that way. Relations between FoE and groups like Road Alert had turned sour. To the NVDA contingent it seemed the regular green lobby had deserted Twyford in the face of the enemy. Were they warriors for real – or just another brick in the wall?

2

Imagining justice

Imagining justice is an essential prelude to achieving it.

– Mahatma Gandhi

During that same summer, while events stubbornly refused to unfold at Twyford, another stand-off was taking place on a much larger scale, at least half a world away. The media spotlight switched to Rio de Janeiro, Brazil, to the United Nations Conference on Environment and Development, UNCED or Earth Summit for short.

UNCED was by far the biggest conference ever convened in the history of international relations. More than 140 national delegations and 100 heads of state from UN member countries were there for the official side of the event, with around 3,000 support staff and at least 1,000 media hangers-on. Somewhere between 15,000 and 20,000 individuals also gathered in Rio, representing other sectors holding a stake in the Summit. Most belonged to the 'independent sector', Rio-speak for non-governmental organisations or NGOs – mainly technical institutes, pressure groups and citizen or consumer groups.

The NGOs had their own meeting and venue, the Global Forum, on a site behind a beach in downtown Rio. The official Summit proceedings were out of the city to the south, at the giant Riocentro conference complex. Green organisations were none too happy about this apartheid, though Rio's taxi drivers made fortunes from hauling delegates to and fro.

UNCED had been conceived partly as a follow-up to another Summit that took place in Stockholm twenty years previously, in 1972, the UN Conference on the Human Environment. At Stockholm, the NGOs had grabbed the agenda and the headlines from the governments, forcing them to agree to far more than they'd bargained for. Was the segregation the UN's none-too-subtle way of ensuring that history didn't repeat itself? This division was not the only aspect of the event and its location that raised independent eyebrows. Rio's city council had spent $14 million on extra policing and other security infrastructure for the event, known locally as Eco '92.

During the Summit fortnight army squads turfed out dope peddlers, street children, muggers, transvestites, pimps and beggars from streets and beaches normally swarming with lowlife. A brand-new $150 million highway, nicknamed the Red Route, was constructed from the airport to the city centre and conference areas, ingeniously avoiding the slums and waste tips that encircled much of the city. Visiting dignitaries were spared the distastefully visible consequences of Brazil's crushing $123 billion foreign debt, the rural poverty that had filled Rio and other cities with desperate environmental refugees.

Despite its indebtedness Brazil was, and is, one of the wealthiest countries in the world. But its wealth lies in the hands of a tiny elite; scarcely 9 per cent of the population owns more than 90 per cent of the nation's land and other resources. Brazil was also famous, or infamous, around the world for its destructive development projects in Amazonia and the ill-treatment of the indigenous rainforest people.

Political nepotism and corruption were rife: even as he greeted delegates like John Major and George Bush to the Summit with all the pomp he could muster, the incumbent President, Collor de Mello, was trying to fend off impeachment proceedings, a fight he was almost inevitably going to lose.

None of this augured too well for the Earth Summit or the Global Forum. Yet many developing country delegations and development interest groups still came to Rio with high hopes.

These hopes were partly buoyed up by the World Commission on Environment and Development's 1988 report, *Our Common Future*, also known as the Brundtland Report after the Commission's Head, Norwegian Prime Minister Gro Harlem Brundtland. This document called for 'a new economic world order' to help equalise the benefits of global trade between the affluent industrial countries of the 'North' and the mainly agricultural and much less affluent 'South': most of Africa, Asia and Latin America.

According to Brundtland and her colleagues, nothing less than a total re-evaluation of global assets, including a just assessment of the environmental as well as the strictly economic costs and benefits of development, could be expected to turn development from its present destructive course. It would set South and North on level terms, using environmental arguments to help, not hinder, the South's economic growth.

Equally, the sheer thrill of having provoked the largest gathering of governments in history was enough to give some NGOs (especially the smaller ones) a sense that something positive was going to emerge from the cauldron of ideas, principles and political deals that was meant to come to the boil in Rio.

Admittedly, the preparatory conferences in Nairobi, Geneva and New York had not been well attended (many groups simply couldn't afford the air fares), but at least the agenda had begun to take shape. Conventions on climate and biodiversity were to be discussed and agreed, the felling of forests regulated and a host of environmental wrongs would be righted.

Alas, the reality was to be something different. Tony Juniper, then the Rainforest Campaigner at Friends of the Earth, remembers all too vividly the first days of the Rio jamboree: 'It was chaos. The Dalai Lama was rubbing shoulders with Roger Moore and a hundred other celebrities, while a weird army of skimpily clad young women and gun-toting security guards ran around trying to organise everyone. If they have cocktail parties in hell then Rio was some kind of earthly manifestation.'

Juniper was among the luckier delegates. Staying with Andrew Lees, the Campaigns Director, and Fiona Weir, the Atmospheric Pollution Campaigner, in a flat loaned by a friend, they were only a few minutes from Riocentro and could get to the heart of the government action while other groups were still stuck on the beach at the Global Forum. Unfortunately, being at the Summit didn't mean that things became any more comprehensible.

The agendas were huge, complex and swathed in detail. Just trying to keep track of a constantly moving merry-go-round of meetings on intellectual property rights, forestry, toxic waste and a hundred other things was a job in itself. Juniper says,

> You had to focus to survive. I restricted myself to just the working groups on forestry but even that seemed to develop into a host of other issues: carbon dioxide fixation by trees and global warming; the rights of indigenous forest dwellers; North–South equity. What started out as one issue soon became fifty and in some ways the forestry debate was a microcosm of what was happening throughout the Summit.
>
> By the end of the Summit the governments agreed the so-called 'Forest Principles' but they were very weak and allowed any government to say that they were following all the principles already.

And simply trying to gain access to the debates was fraught with difficulty. 'We were excluded from a lot of the meetings and press conferences,' recalls Fiona Weir. 'We spent a lot of time sneaking into places and trying to look inconspicuous for as long as possible before someone spotted us and had us thrown out.'

And if well-informed campaigners were having trouble making sense of the proceedings, the media's task was even harder. 'The press had a tough time getting a handle on what was really happening,' says Weir. 'They couldn't cope with the complexity of the issues and didn't really have a grasp of the context. Inevitably, the journalists looked for soundbites and the coverage collapsed into clichés.'

As confusion reigned it became obvious to most groups that they were unlikely to have any impact on the conference. Frustration dictated new tactics. The FoE team inflated a huge model chainsaw on the roadside just a few hundred metres from the conference centre and had the satisfaction of watching every head of state have to drive past it.

When they didn't have a handy inflatable, they targeted the media instead. Draft agreements and position papers leaked by sympathetic government delegations to the FoE team became a prime source of news stories and Andrew Lees – never the most laid-back of campaigners – went into overdrive and churned out press releases at a rate which staggered his colleagues.

'Andrew was unstoppable,' remembers Juniper. 'He just kept knocking them out and persuading journalists to write about them. Even at the end of the conference he was still printing off press releases as they removed the furniture around him. It was like Captain Ahab had discovered the photocopier.'

When Lees wasn't swamping the media with stories of governmental duplicity he was spending his time networking, lobbying and even building his own theatrical props. Juniper recalls that Lees decided to forgo sleep one night and spend his time manufacturing a large placard with a skull and crossbones emblazoned at the top and the words 'toxic information' crudely drawn below. The next day Lees disappeared off to the Global Forum on the beach and tried to disrupt a corporate press conference only to be forcibly ejected by some heavily armed security guards.

When the Summit was over and the last government delegation had departed, it was time to take stock. The conference centre lay eerily empty while the Global Forum site was now just an abandoned ghost town of litter, populated by a few bewildered indigenous people, crystal healers and drunks wondering how they were going to get home.

And the cruellest blow of all was yet to fall. Only a short time after the Rio Summit closed the world was plunged into economic recession. It was perhaps one of those inevitable ironies that at exactly the moment when the world realised that a massive financial input would be needed to tackle environmental problems, the global economy went into a violent dip. Promises of international largesse evaporated as the industrialised world struggled with a sharp economic downturn. What could have been a defining moment in history became simply a 'what if?'.

But despite the disappointments, many greens remain sanguine about the event. Says Juniper,

> We knew that the conference would never live up to the hype but it was still an important event. It didn't make things better or worse but it did make things different and the reverberations are still with us – both good and bad.
>
> I think the most important lesson was the underlying reasons why governments felt they had to attend in the first place. Public concern about the environment had forced politicians to take the issues seriously and make some kind of gesture. The downside of course is that Rio became exactly that – a gesture. The whole thing was really a gimmick to satisfy public opinion and unfortunately it was all very effective. All of the issues that concerned the public then are still with us: habitat destruction, climate change, pollution. But they are all much lower profile than before.
>
> On the other hand, if public opinion can make politicans go to Rio then ultimately it might force them to make some changes. For pressure groups like Friends of the Earth, mobilising the public will always be a

Figure 2 'Rio didn't make things better or worse but it did make things different and the reverberations are still with us – both good and bad' © Dylan Garcia

much more effective tool for change than spending our lives in worthy subcommittees poring over endless redrafts of conventions that will never be implemented.

And a sharp lesson in tactics was not the only booty that environment groups could take away from Rio. Two conventions had come into existence, covering climate change and biodiversity.

The Climate Convention was admittedly short on real targets, the international energy and automobile lobby had seen to that. But it did at least require governments to 'aim to return' their emissions of greenhouse gases to 1990 levels while drawing up inventories, impact assessments, country reports and other documents.

None the less, even these rather woolly requirements proved too much for the USA – which had played a lead role in the creation of the convention – and President Bush flew home to his re-election campaign proudly announcing his refusal to sign as further evidence of his presidential credentials. It was left to his nemesis, President Clinton, to have the honour of signing the Climate Convention on behalf of one of the world's most polluting countries.

Similarly, the Biodiversity Convention was unlikely to cause even the most laggardly governments much trouble. But neither was it entirely useless. The UK Government, for instance, has recently devised a national action plan as part of its convention commitments. This sets real numerical

Figure 3 Friends of the Earth-ly reminder: Earth Summit poster, Westminster underground station © Patrick Sutherland/FoE

conservation targets for the first time, a powerful tool for groups campaigning for wildlife.

Equally, Agenda 21 – the 'action list' for sustainable development formulated by the Rio participants – may not be much honoured globally but is proving a useful tool at the local level. Local FoE groups throughout the country are badgering their councils to turn Agenda 21 into a real set of priorities for grassroots action and many are making significant progress.

And much else that was good also came out of the Summit. Juniper thinks that Rio may have been a turning point in cooperation between groups from the North and organisations based in the less developed countries of the South:

> For all its failures, Rio really did open our eyes for the first time to what the southern groups were doing. Issues like equity were discussed in detail and we found a common agenda so that we're now much more unified about where we want the world to go.
>
> Ideas like 'ecological space' – the principle that every human has a right to an exactly equal portion of the Earth's resources and no more – are now central to our campaigning work and are fully endorsed by all of our partners in Friends of the Earth International and beyond. We probably wouldn't have achieved this kind of international unity so quickly if we hadn't all met in Rio.

'It was also the first time,' adds Weir, 'that we saw a whole range of linked issues such as the environment, the economy, jobs, human rights and social equity being considered together. It's true that Rio represented something that had already happened rather than being something new in itself, but that shouldn't detract from its overall significance.' Juniper concludes:

> At the end of the day, the Rio Summit didn't do much for the planet. But it taught Friends of the Earth that giving governments grief is still the best way to achieve a better world. And just as importantly, it taught us that we weren't alone – we had to think and act globally along with our partners in other countries and base our campaigns on the concerns of all of us. After 1992, thinking and acting globally became a reality, not just a slogan.

Back on the M3 construction site at Twyford Down, the growing band of protesters must have heard the news from Rio about global curbs on greenhouse gas emissions and wondered when reason and justice would make some sort of appearance in rural Hampshire.

When the final moment of truth came at Twyford at the start of December, it came suddenly. The protesters were behaving peaceably and nothing much seemed to be happening. Then squads of police and private security men appeared out of the blue, beat them up and threw them out. Many were arrested. Four went to jail.

The new wave of public upset over road-building continued into 1993 and beyond. The extension of the M11 through houses in Wanstead in 1994 widened the scale and intensity of public protest and of its repression by police and hired security thugs. Roger Higman reflects:

> I think the Government could have steamrollered over the Direct Action people the way they did over the miners, if it wasn't for the fact that they had a lot of support from a lot of Conservatives. The Tories were frightened by the obvious alliance between people who were at best on the margins of society and well-heeled people who were used to getting their own way. And of course John Major had a much smaller majority than Thatcher had, so he was far more vulnerable to backbench pressure.

The Direct Action outfits came from very indeterminate political roots and an ideology centred on ideas of natural rather than social justice. Roger Higman admits that 'it went through our minds that Earth First! might be the Friends of the Earth of the year 2010'.

3

Destroyers of the temple

> We seek a renewed stirring of love for the Earth; we urge that what people are capable of doing to the Earth is not always what they ought to do; and we plead that all people, here, now, determine that a wide, spacious, untrammelled freedom shall remain as living testimony that this generation, our own, had love for the next.
>
> – David Brower

In the USA and Britain, most of the national institutions commonly seen as part of a classic pro-nature tradition were up and running long before 1960. Their main focus for many years had been preserving particular landscapes, wild species or the rural scene in general – the countryside – from destructive misuse. Some also sought to make public access to the wild both secure and manageable.

In Britain the National Trust, born in the late nineteenth century as a reaction against the railway construction boom of the Victorian era, especially the effect of railway cuttings on beauty spots like the Lake District, was one of the oldest of these bodies. The Royal Society for the Protection of Birds, the largest of the more wildlife-oriented groups, had been set up in 1888 by a group of well-heeled women in Altrincham protesting against the use of wild bird feathers in the hat trade. The Fauna and Flora Protection Society began life in 1903 as the Society for the Protection of the Fauna of the Empire. The Council for the Preservation of Rural England (CPRE), started in 1926, spearheaded a movement to demarcate National Parks in areas like the Cairngorms and the South Downs. Part of the idea was to emulate the protected areas system then growing fast in the USA, but CPRE was also concerned about the plight of industrial workers, following a succession of failed attempts by parliamentarians to guarantee public access to open countryside. In the 1940s and 1950s, a scattering of County Naturalists' Trusts were founded and later coalesced under the Society for Promotion of Nature Reserves, now the Wildlife Trusts Partnership.

Though many of these bodies or networks had roots in radical, populist or liberal ideals, most had settled into a comfortable rut by the 1940s and 1950s. Typically, their membership consisted of well-meaning middle or lower middle class ramblers, country-house buffs and hobby naturalists.

A more scientific wave appeared with the foundation of a British Ecological Society in the early 1940s. Sir Julian Huxley, Arthur Tansley and other eminent scientists argued for an official protected areas and wildlife protection system to preserve the scientific research potential of wild sites as much as amenity or heritage values. To look after these new reserves it was necessary to develop new science-based knowledge and skills in environmental management. The high esteem in which a scientific approach to problem-solving was held in British political culture at the time won the day.

A sympathetic Treasury official, Max Nicholson, summed up the new spirit thus: 'Nature Reserves are regarded perhaps more than anything else as outdoor laboratories where the workings of nature can be studied . . . Paradoxically we can ensure the survival of wild places in Britain only by finding out what happens when we interfere with them.' On this basis, most of the foundations of the official side of nature conservation in Britain were laid during the first flush of 1940s postwar reconstruction. The legal keystone was the National Parks and Access to the Countryside Act, passed in 1949. With that came the first agencies for nature stewardship, the Nature Conservancy Council (now English Nature, Scottish Natural Heritage and the Countryside Council for Wales) and the National Parks Commission.

Although official recognition helped the cause of nature conservation in many ways, it left the voluntary movement feeling at a loss. Nicholson described their condition in the 1950s and 1960s as typified by low morale, weak leadership, frail finances and elderly memberships. As for the new legislation, though it paved the way for the creation of National Nature Reserves, National Parks and (by 1962) over 1,700 Sites of Special Scientific Interest, it also prematurely crystallised the debate about what conservation was truly *for*.

More was happening on the international scene. A consortium of national scientific or science-based nature conservation bodies was formed in 1948 at Fontainebleau, near Paris, the UIPN or Union Internationale pour la Protection de la Nature. It was conceived as a UN for those who cared about the future of the natural world and it echoed a general enthusiasm for League-of-Nations type internationalism in the wake of two traumatic world wars. Its membership included entire sovereign states as well as voluntary bodies, a unique mix. In the 1950s the 'protection' part of its name became changed to 'conservation', making it the UICN (in English, IUCN). The change was made to avoid the unwanted imputation that the alliance was interested only in fencing nature off behind barbed wire to remove from it all taint of human presence.

Conservation of natural resources, linked to the 'wise use' concept, had instead become the watchword. French-speaking members insisted that the word 'conservation' meant the wrong thing in French, implying the preservation of jams, pickles and potted meat! But their objections were overruled, the organisation having by now taken on a more Anglo-Saxon cast of members, including the Conservation Foundation (also founded in 1948) and other wealthy and vocal US groups. It was unsurprising that to a large extent they led the 'wise use' debate in this international arena, for it was a constant source of factional in-fighting for most of them at home.

Back in the USA, the classic pro-nature lobby was dominated by the Sierra Club (formed 1892) and National Audubon Society, dating from 1896. The founders of both were mainly affluent, leisured 'outdoorsmen' hooked on fishing, hunting or camping. In his recent overview of environmental idealism in America, *Losing Ground* (1995), journalist Mark Dowie says:

> Members of the new organisations were mostly well-bred [and] virtually all were white Anglo-Saxons, an ethnic group that would dominate American environmentalism until the late twentieth century. The gospel of the hour was efficiency, the management of resources the new science. The notion that natural resources belonged to all people was debated, eventually accepted and written into laws whose interpretations are still being challenged in court.

This shift ran decidedly counter to the fledgling conservation movement's original ideas and loyalties, represented by 'transcendentalist' poets and writers like Rousseau, Emerson and Thoreau, together with Thomas Cole and others in the 'Hudson River School' of nature painters. Their concern over an imminent collision between timeless nature and the modern forces of industrialisation and urbanisation had obvious counterparts in England in the works of Wordsworth, Turner and Ruskin. But in America the idea of a nature at risk from human abuses struck home with more instant force than in Europe.

Huge areas of forest had been cut down within living memory to create farmlands and build settlements, highways and railroads. Wildlife had disappeared along with the wilds: the entire beaver population of New England had been exterminated by fur traders by 1800 and the innumerable bison of the Great Plains had been almost wiped out by hunters by the late 1880s. Says Dowie, early settlers like Daniel Boone and the great painter of birdlife, John James Audubon, 'had seen in the grandeur of North America many things worth protecting from the growing environmental degradation of industrial Europe.'

In his *Wilderness And Plenty* lectures of 1969, pioneer British conservationist Frank Frazer-Darling was to muse: 'The onslaught of the nineteenth century on the forests of North America was so shocking I have the feeling that that was the reason for the early rise of a sense of conservation there.' Evidence for this rise was the foundation in 1872 of the world's first National Park in the modern sense, Yellowstone, created mainly with a view to preserving a residual herd of buffalo thought to linger there. Another advance came around 1905 when the US Forest Service was founded in the Department of Agriculture. The recently elected Franklin D. Roosevelt sanctioned this step on the advice of a young forester called Gifford Pinchot.

Pinchot was fearful that the historic loss of forests he had witnessed during travels in Europe was about to afflict the American forests, too. He had also seen modern forest management practices applied in Germany to deal with the problem, and wanted to see those principles put into practice in America *before* the 'American Colossus' of unbridled growth driven by capital investment took over. The exploiter was, said Pinchot, 'fiercely intent on appropriating and expropriating the riches of the richest of all continents – grasping with both hands, reaping where he had not sown, wasting what he thought would last forever.'

Pinchot belonged to the exclusive Boone and Crockett hunters' club, where he made friends with Roosevelt during the latter's youth. When another Roosevelt, Theodore, came to power, Pinchot, by now in charge of a Forestry Division in the Department of Agriculture, persuaded him that business tycoons with scant concern for nature were in control of America's least replaceable natural assets, and needed to be curbed.

It was Pinchot who coined the term 'wise use', to describe an accommodation between nature preservation interests and the economic imperatives to extract natural resources like timber or minerals or to clear land. In the forest context, it meant taking areas of forested wilderness into public ownership and creating a federal bureau to protect them from purely destructive use.

John Muir dismissed the 'conservationist' approach as too full of loopholes to allow the country's wilderness treasures to survive unmolested by 'devotees of ravaging commercialism' who 'instead of lifting their eyes to the God of the Mountains, lift them to the Almighty Dollar'. He too lobbied Roosevelt with his opposing 'preservationist' viewpoint, meeting him in Yosemite Valley in 1903 to demonstrate his point on the spot.

Muir told the President that everything he saw around him should be protected by legislation as 'a cathedral where people could experience the wild in their souls and rejuvenate themselves after months in the urban jungles of America', writes Dowie. Roosevelt must have liked the idea. A few months later he introduced a federal bill to protect Yosemite Valley and its surrounding peaks, already a national park, from any form of private exploitation or environmental damage.

Muir's model among nineteenth-century literary nature lovers was Henry David Thoreau, who lived out the life of a wilderness hermit, was the author of a book called *Civil Disobedience* and had served time in jail for refusing to pay taxes. When the passage of shad, a migratory fish, was threatened by dam-construction schemes on the Concord and Merrimack Rivers, Thoreau voiced eco-saboteur sentiments that now sound curiously modern. 'Who hears the fishes when they cry? I for one,' he assured his finny friends, 'am with thee. Who knows what may avail a crowbar against the Billerica Dam?'

Pinchot and Muir inevitably took different sides in 1912 over the damming of the Hetch Hetchy Valley wilderness site, to provide drinking water for San Francisco, some 200 miles away. 'These temple destroyers ... seem to have a perfect contempt for Nature', said Muir. But Pinchot was 'fully persuaded that by substituting a lake for the present swampy floor of the valley, the injury is unimportant compared to the benefits ...'

Muir's view of the living world, though preservationist, was very much in tune with ecological 'whole systems' or holistic thinking as we acknowledge it today. He once remarked: 'When we try to pick out anything by itself, we find it is hitched to everything else in the universe.' He didn't trust the developers anyway, but his main objection to pursuing 'wise use' compromises was that we probably weren't wise enough to make them work even with the best will in the world.

David Brower, founder of Friends of the Earth, was born close to Hetch Hetchy in 1912, the same year that Muir fought his last, losing battle there. As a child, Brower often played beside the lake created by the dam. He would later count himself proudly among Muir's and Thoreau's most loyal spiritual descendants. The young Brower dropped out of college when he was 19, around the time of the Great Depression, and disappeared into the Sierra Nevada.

He emerged first as a sought-after climbing instructor then as a feisty pro-nature lobbyist, defending mountain ranges and other wilderness areas. To do this he had to operate in Washington's and New York's corridors of power, the last place he wanted to be. But where else would his influence count?

'Mining exceptions' to the preservation of protected mountain areas, logging rights permitted in vast areas of the National Forests: slack technicalities like these became Brower's battleground and as a rising star in the Sierra Club he relentlessly harried those who used them for ulterior purposes. In 1960 Brower helped put together an exhibition in Yosemite Valley of photographs of the place by Ansell Adams and other eminent photographers, with accompanying texts by writers like Muir and Thoreau. This he later reassembled in book form, the first of a long and lucrative series of $125 coffee-table conservation picture books to be published by the Sierra Club.

Within nine years, 10 million dollars' worth of these 'Exhibit-Format' books had been sold. Building on this new-found wealth and the awareness the books also raised, the Club expanded, with Brower now its Director, from a membership of 7,000 to 77,000 It also became a potent political lobby, influencing legislation on a wide range of matters. Brower was personally involved in all these campaigns and was a primary force in advancing a national Wilderness Act, which remains in force to this day.

He was also held responsible by Government mandarins for stopping plans to construct dams, including two intended for the Grand Canyon. Feeble attempts had been made to justify these schemes by arguing that it would be easier for visitors to view the Canyon's walls from boats touring the lakeland that the dam barriers would create. Brower ran full-page advertisements in the national press that screamed: 'WILL THEY FLOOD ST PETER'S SO TOURISTS CAN SEE THE ROOF BETTER?!' In his 1971 biography of Brower, *Encounters with the Archdruid*, John McPhee says: 'To his foes in the Department of the Interior and Bureau of Reclamation, David Brower is the Antichrist.'

Within the Sierra Club, and increasingly outside it, those of the preservationist or pro-Muir persuasion became known to their conservationist detractors as the Druids, a label they decided to wear with pride. David Brower was seen as their chief protagonist. As disputes between the rival factions intensified, Brower grew more and more convinced that the conservationist agenda had become hijacked by 'wise use' interests that were indistinguishable from Muir's 'temple destroyers'. In 1969, he quit the Club.

Two lines of reasoning prompted Brower's drastic action. One was the thought that the pro-nature mainstream groups like the Sierra Club and National Audubon Society had been seduced away from their original intentions by the argument that campaigns for conservation or against destructive development should stem from realistic, science-based principles, from 'thinking objectively' about the pros and cons before adopting a position. Brower once set on record the view that 'thinking objectively is the greatest threat to Nature in America'.

For him there had to be scope for intuitive responses to the issues, stemming from a subjective bond formed between people and wild places. This conviction led Brower to found a new association named after his hero, the John Muir Institute for Environmental Studies. But Brower had also realised from his seventeen years' experience of running the Sierra Club that the whole ground of the debate had shifted, and with it the basic constituency wanting to become engaged in living world issues.

He figured that there might be a new and far larger constituency out there, whose members were on average around half the age of the old guard. To cater to this shadowy band, he conceived a new type of pressure group, a group which from the start he intended to clone and multiply in other countries around the world. The new organisation would be a global Druid bodyguard, keeping the 'wise use' weasels and rank destroyers out of nature's temples. Brower's wife, Anne, named it Friends of the Earth.

Nothing was more indicative of the shift of ground Brower had sensed than the appearance in 1962 of *Silent Spring* by Rachel Carson. More than any other single document, it established the term and the idea 'environment' in common usage. The book drew dramatic linkages between the saturation use of chemicals like DDT in intensive modern farming, and the suspicious scarcity of songbirds like the American robin on the rural, suburban and small-town scene.

Rachel Carson was a self-taught marine biologist (she later graduated formally in the subject) living in Silver Spring, Maryland, not far from Washington DC. Her other published works included an inspirational book *The Sea* about linkages between human life and life in the oceans. But *Silent Spring* was the book that first put captains of industry and big business on the rack, tracing a toxic pollution threat to human health, occupational safety and the natural world. She wrote,

> The most alarming of man's assaults upon the environment, is the contamination of air, earth, rivers and sea with dangerous and even lethal materials. For the first time in the history of the world, every human being is now subjected to contact with dangerous chemicals, from conception until death. The question is whether any civilisation can wage such relentless war on life without destroying itself and without losing the right to be called civilised.

For the rest of her life, which was tragically cut short by cancer, Carson was vilified by the hired stooges of the multi-billion-dollar agrochemicals industry. It was a sign that her criticisms had found their mark, though some admittedly did turn out in the long run to be slightly wide of it.

To try and counter the impact of *Silent Spring*, the industry launched a propaganda counter-offensive, especially in developing regions, where it saw its most promising market growth area. Carson and environmental activists in general were accused by industry-funded groups like the Groupement International des Fabricants de Produits Agricoles (GIFAP), based in Belgium and funded by the 'Big Six' transnational pesticide corporations, of plotting mass murder by planning to deprive starving millions in the Third World of the means to save their crops from pests and diseases. Despite such rearguard actions, DDT-based pesticides were banned in most countries within ten years of *Silent Spring*'s launch. For the first time, a major industry had been identified and very successfully impeached as an abuser of the environment.

Though not limited to wild nature, Carson's work perhaps provided a link between preservationist and conservationist points of view. It indicted industry's assault on nature in terms of impassioned outrage, yet justified restrictions on these abuses in terms of health, efficiency and safety as well. Notions of offended justice were accompanied by mild suggestions as to better ways to achieve industrial progress without degrading people's surroundings, exhausting natural resources or creating streams of hazardous waste products.

Carson's middle way reflected her empathy with another founding father of nature conservation in America, Aldo Leopold. In his master work *A Sand County Almanac*, Leopold propounded an ecological view of the world that 'simply enlarges the boundary of community to include soils, water, plants and animals'. Within this community, people were simple members, not custodians or commanders. In Leopold's ethic, 'a thing is right when it tends to preserve the integrity and stability and beauty of the biotic community. It is wrong when it tends otherwise.' For Carson, DDT was an obvious wrong.

The early 1960s also saw the appearance of Barry Commoner's *The Closing Circle*, a more general indictment of the polluting habits of large industries, including the unsafe storage and disposal of hazardous wastes. Other key works of the time included Paul Ehrlich's *The Population Bomb*, which warned of global nemesis unless skyrocketing birth rates in the world's poorest nations were curbed or the new mouths provided for.

A cascade of new environmental legislation appeared in the wake of this wave of concern, including the Wilderness Act (passed in 1964), the first Clean Water Act (1965), the Clean Air Act (1967) and the Wild and Scenic Rivers Act of 1968. On the unofficial side, the Environmental Defense Fund came into being in 1967, regarded by some historians as the first of a new breed of national environmental organisations in America.

Mark Dowie points out that although the labels 'environment' and 'environmentalism' date mainly from 1962 and *Silent Spring*, the American environmental movement did not emerge ready-made during the 1960s. A vast tradition existed before Carson set pen to paper. 'What became known as environmentalism was an amalgam of resource conservation, wilderness preservation, public health reform, population control, ecology, energy conservation, anti-pollution regulation and occupational health campaigns.'

A pedantic insistence that environmental ideas as they emerged from the 1960s were only remoulds of existing notions, has many adherents. Some have looked back as far as classical and biblical times and claimed to find the same old prophecies of doom, the same concern for wild creatures and places, the same anxiety over the state of the land and its resources that is expressed in much of the environmental literature of the 1960s and 1970s. Others feel environmentalism has no true precedent.

Considering the broader historical and social context of the 1960s, they have a point. Consumer culture had changed the lives of people in the North almost beyond recognition since the 1940s, filling their homes with new products and new materials, all using up more and more energy in various forms, and producing more and more waste and pollution in the process.

Despite this bonanza of possessions and comforts, life was not secure. Nuclear destruction was an ever-present threat. The Cold War born of the arms race flared up occasionally to remind everyone of it, most notably during the Cuban missile crisis of late 1962. Around the developing world, processes of decolonialisation were in full swing and new loyalties were being courted across the Cold War's invisible lines.

In 1956 the French political theorist Alfred Sauvy described the emergence of a Third World geopolitical bloc with an agenda and destiny of its own, which would in time become the leading world power. Sauvy used the term Third World to imply a parallel with the Third Estate, the mass of citizens without privilege who had risen up during the French Revolution to seize power and glory from their erstwhile masters.

The 'wind of change' that Harold Macmillan described as blowing through Africa was soon blowing through America, too, in the civil rights movement led by figures like Martin Luther King. The demonstrations and civil disobedience campaigns of the black emancipation movement typified a mood of popular revolt amid apparent affluence that became the decade's trademark.

Science was undergoing a changing of mental gears that Thomas Kuhn referred to in 1962 as a 'paradigm shift', a transformation in how people think of the world. With space flight now a reality, thoughts turned amongst other things to what polymath and technophile Buckminster Fuller called 'Spaceship Earth'. David Brower later summed this notion up in his own, less mechanistic way as: '. . . the sudden insight from Apollo. There it is. That's all it is. We see through the eyes of the astronauts how fragile our life is, how thin is the epithelium of the atmosphere.' Way before Apollo, Brower had incorporated this idea into the FoE name.

The influence and scope of the mass media was changing, too, especially through the miraculous yet invasive presence of the televised image in millions of homes. Not only could this medium reach audiences as never before, it could also reach and reveal subject matter never before made readily accessible to the masses, including images of events like the escalating conflict in Vietnam which many might prefer to ignore.

As well as reflecting huge social and cultural changes, television also bred an advertising culture of its own, promoting not just goods and services but lifestyles and attitudes, too. Above all, the position in society of young people of college and high school age was changing. Identified by advertisers as a lucrative new market during the 1950s, the teenager had grown up by the 1960s into an assured and outspoken participant in a youth culture creating its own values, standards and politics.

Released from parental apron strings by a perceived 'generation gap' of vast proportions, the 'we generation' of the 1960s responded to the maelstrom of change around it with gusto, enjoying an atmosphere of revolution and experiment that poet Brian Patten has described as 'a fizzly electrical storm'. Depending on temperament or chance as much as anything, there was a choice of at least two revolutions for the taking.

One was the 'tune in, turn on, drop out' anarchic hippydom commended by Dr Timothy Leary and other advocates of chemical bliss. The other, more evident towards the latter half of the decade, tended towards militant leftist ideals with warrior overtones, the anarchy of Baader–Meinhof and the Red Brigades.

As Ian MacDonald suggests in his book about the Beatles' records and the 1960s, *Revolution in the Head* (1994), recent right-wing attacks on 1960s' youth culture are oddly self-contradictory. These critics blame the decline in economic and educational achievement during the 1980s, along with rising rates of assault, rape, robbery, drug abuse, divorce and abortion in the 1990s, on the permissive society of the 1960s, supposedly fomented at the time by left-wing subversives and anarchists. According to MacDonald,

> [The chaos] was created neither by the hippies (who wanted us all to be together) nor by the New Left radicals, all of whom were socialists of some description. So far as anything in the Sixties can be blamed for the demise of the compound entity of society, it was the natural desire of the 'masses' to lead easier, pleasanter lives, own their own homes, follow their own fancies and move out of the communal collective. What mass society unconsciously began in the Sixties, Thatcher and Reagan raised to the level of ideology in the Eighties: the materialistic individualisation – and total fragmentation – of Western society.

MacDonald links this scary cultural implosion to the 1960s' consumer boom in labour-saving devices. The countercultures usually blamed for this were, in fact, resisting it. 'Far from adding to this fragmentation, they aimed to replace it with a new social order based on either love-and-peace or a vague anarchistic European version of revolutionary Maoism.'

For those who lived through the 1960s as a product of the Baby Boom (or its equivalent in Britain, where it was more lugubriously termed the Postwar Bulge), Ian MacDonald's analysis has a clear and penetrating ring of truth to it.

And yet, surprisingly, MacDonald leaves one key factor out of his equation. It is the genesis during the 1960s and early 1970s of pro-environment pressure groups, mainly in North America and northern Europe, key strongholds of consumer materialism. Did this not count as an important counterpoint to the forces of fragmentation he identifies, a vital thread of utopian and libertarian thinking? And hasn't that pro-environment thinking survived in one form or another through succeeding decades, not unlike the music of its original era?

It was only a shuffle of feet from Ban the Bomb marches or demonstrations against the Vietnam War to concerns over other global life threats, and only another short step back into an obsession with personal health and lifestyle and thence to concern over environmental life-support systems.

Even as 1960s' youth culture took environmental thinking to heart, it was also changing it, using its ideas and expressions to give shape and movement to others. It seized on the notion of an ecosystem, for example, as a neat metaphor for the kinds of integrated and equitable political, social and economic systems the times seemed to be calling for. If this was nothing new, the word 'new' must mean nothing.

4

Creature context

> There is only one being in possession of weapons which do not grow on his body and of whose working plan, therefore, the instincts of his species know nothing and in the use of which he has no correspondingly adequate inhibition. That being is Man. We must build up those inhibitions purposefully, for we cannot rely upon our instincts.
>
> – Konrad Lorenz, *King Solomon's Ring*, 1952

If environmental thinking after 1960 was one of the vital contemporary 'new paradigms' Kuhn had spoken of, that did not have to mean that it owed nothing to what went before. Far from it: the nature of new paradigms is not only to modify ideas or values but also to be modified by them. The deep-seated traditions of classical natural history and (mainly) nineteenth-century ideals of animal welfare, fused in the 1960s and shortly before, with twentieth-century notions of species survival and a continuing fallout of anxiety neurosis from World War II and the Cold War, to create a whole new nexus of concern, wanting only a new medium to give it form and growth.

During the later 1950s and early 1960s, one of the most popular British TV programmes was a natural history slot called *Look!*, presented by the popular and amiable naturalist Peter Scott, son of the explorer Captain Scott, hero of Antarctica. Peter Scott's interest in wildlife had begun as an enthusiasm for shooting birds rather than filming or protecting them. Partly out of deference to his father's dying wish that he should dedicate his life to natural history, Scott gave up hunting and in the 1950s founded a world-famous wildfowl reserve at Slimbridge, in Gloucestershire. He was a natural choice to present *Look!* The films he introduced on the BBC programme, to an audience mainly composed of older children, became dependable crowd-pleasers but were limited to big game spectacles and the odd underwater adventure, often recorded during specimen-hunting expeditions funded by zoos.

In 1956, Scott's programme included a film just 20 minutes long by a previously obscure German cinematographer, Heinz Sielmann. It showed the 'secret life' of a family of woodpeckers, filmed with astonishing intimacy through the back of their nesting hole and revealing aspects of the bird's parenting and breeding behaviour never before observed by scientists, let alone by a viewing public. The feedback from this screening caught the show's producer, Desmond Hawkins, by surprise. 'It was the first time,' he says, 'that the BBC's switchboard had been jammed by viewers' calls, apart from for current affairs programmes. They all wanted more films like it. We were amazed.' There was a simple explanation for this response, however: nature had just become a 'current affair'.

Max Nicholson, by now director of the Nature Conservancy Council, thought he knew what lay behind this surge of interest. The canny Nicholson had just set up a committee which included Sir Julian Huxley (by now head of Unesco), an advertising executive named Guy Mountfort and an émigré Czech business tycoon Victor Stolan, to establish a new kind of organisation to raise money for international conservation project work. Mountfort came up with the name World Wild Life Fund for this body. The name was later amended to World Wildlife Fund (WWF) to avoid connotations of strip joints and nightclubs.

Nicholson's concept was to use PR skills borrowed from Madison Avenue and the pull of the mass media to rally the public, the business world and their money and loyalties behind nature conservation. As Fred Pearce recounts in his 1991 book on the environmental revolution, *Green Warriors*, Max Nicholson didn't think much of natural history TV as a vehicle for this manoeuvre, and told Scott so. 'I argued with him [Nicholson told Pearce] that his TV programmes were frankly escapist . . . that these birds that they record might not exist in a hundred years' time if they didn't draw attention to the environmental problems they faced.'

This was an unjust criticism in the case of Heinz Sielmann's work. Though Nicholson could be fully excused for not knowing it, Sielmann's agenda as a cineaste was to some extent determined by the pacifist theories of the man who practically invented modern animal behaviour studies, Konrad Lorenz, author of *King Solomon's Ring*. Sielmann's studio was a stone's throw from the Institute where Lorenz conducted his lab studies – and from the Arriflex factory, where the first lightweight and truly portable TV camera was being developed. The twenty or so major natural history film projects that arose from this fortunate combination of circumstances were to bring state-of-the-art life science knowledge, as it emerged, to the general lay public.

A former Luftwaffe colleague, a young artist named Joseph Beuys, worked with Sielmann on several of these projects. Working on the film locations was a therapeutic process for the shell-shocked Beuys and many of his mature art projects recall hours and days spent observing animals from tiny hides, waiting to capture moments of truth and action. In later life, Beuys was also to join with Petra Kelly and others to help found the German Green Party.

The founders of WWF could have had no inkling of these converging destinies back in 1961, however, when Nicholson persuaded Scott to come to the headquarters of IUCN in Morges, Switzerland, to meet the other members of his committee and become their Chairman. Legend relates that Scott, having agreed to join the group, sketched the Fund's panda logo on a restaurant paper napkin. If the story is true, it would make that napkin arguably one of the most valuable in history.

The panda provided WWF with what PR and advertising executives would call a 'corporate identity'. It also worked as a means of franchising the WWF message through a global network of national chapters, a key development in Nicholson's view. Keeping these national groups in unified corporate order should be easy: if they stepped out of line, they'd risk their franchise on the panda.

In the beginning, Nicholson and his colleagues had it in mind to offer their new organisation to IUCN as its fund-raising arm. The Union's scientific standing was imposing. Its expert commissions could advise on how best to spend the funds raised. It also had a worldwide membership of national organisations that offered ready-made task forces for implementing conservation projects. And yet soon after this suggestion arose, the group got cold feet and opted for a semi-detached relationship with IUCN rather than full partnership under one management.

It was felt that the IUCN style of quasi-governmental consensus and scientific hair-splitting might cramp WWF's corporate style or limit its PR and policy options. So an arrangement was made whereby IUCN would provide strategic advice and network project activities as and when required, in return for a core subsidy from WWF to help the Union maintain an international presence as an inventor and promoter of new environment laws, protected areas networks, research initiatives and the like.

The WWF national organisations were to stand alone as fund-raising, awareness-raising and educational bodies in their own right, channelling part of their income to a world headquarters next door to IUCN. But WWF International's running costs would not be drawn from this income; all of it would go direct into projects. The running costs would come from a club of anonymous philanthropists, the 1001 Club.

Using their special aptitude for turning the persuasive power of the media and powerful business connections to high-profile advantage, the national organisations and WWF International soon became – or at least many in IUCN's old guard felt they became – the tail that wagged the Union dog. WWF's founders did not escape a taste of this medicine themselves, however.

Nicholson had been especially keen for the organisation to go global from a European rather than a US base. Part of the reason may have been an anti-American feeling that many of his generation shared. He may also have bridled at the way environmental groups in the USA seemed to be overrun by 'Wise Users', or he may have felt that America might not understand the mixed blessings of postwar reconstruction, not having been a theatre of war. Whatever his reasons, he got his wish and the satisfaction of seeing WWF-US formed as an offshoot of a Europe-based WWF International, rather than the other way around. But the US chapter 'always tried to go its own way', Nicholson groused. 'The Americans never saw the point in a global organisation.'

A strong motivation behind WWF's creation was a sense of deep disquiet about the fate of the animal kingdom in places like East Africa as these wildlife-rich regions became decolonised. Nicholson argued that tourism in National Parks like the Serengeti would bring Africa's new nations far more foreign revenues than they could ever get by digging them up and cultivating coffee or raising cattle in their place. Meeting Tanzania's Julius Nyerere and forty-nine other likely future African leaders at Arusha, Nicholson helped lay the basis for a conservation ethos for independent Africa, mainly based on economic self-interest and the prediction that wildlife tourism could be worth £100 million to Africa.

For Scott and for quite a few other luminaries in the nature conservation movement, the notion of putting a price on Nature's head went against the grain of purely ethical, poetic arguments for saving creatures like the rhinoceros from extinction. 'He [Scott] became keen on the idea that we had no right to exterminate animals. WWF always concentrated on losses to mankind, partly because we were already regarded in some quarters as nutcases,' Nicholson later explained. He was referring to the abuse that had begun to creep into media coverage of conservation issues, labelling those who promoted them as weirdos in sandals, muesli eaters and so on.

Inside WWF, there grew a consciousness of two types of supporter that the organisation should cater to. One was the hard-nosed sceptic who wanted evidence of practical problem solving, firmly based in hard science or economics. The other was characterised by adman and WWF Trustee David Ogilvy as a 'little old lady in tennis shoes' worried about furry animals.

It is quite possible that neither of these stereotypes ever existed in a pure form. But it was certainly true that saving wildlife involved a number of very different and not always compatible motivations or philosophies.

The Wise Users believed that science and economics could come up with 'objective' reasons for saving animals and plants from extinction. The need to protect species was made explicit in a metaphor that owed much to Buckminster Fuller's 'Spaceship Earth', the popping rivets analogy. If a plane up in the air loses a rivet or two it will keep flying, just as the Earth might afford to part company with a species or two. But at some point, nobody could be sure when, the rivet will pop that will cause the whole craft to break up in mid-air. Saving species was, by this analogy, self-defence.

Worldlier ideas, such as the potential value of undiscovered medicines and other economic resources hidden away in rainforests and other natural habitats, later became more fashionable than popping rivets as a baseline argument. But the gist of the 'wise use' attitude was that science equalled plain common sense, cost-benefit accountancy equalled objectivity.

Animal welfare was a different, much older story, closely connected to wider cultural and religious, or liberal and humanitarian, traditions of respect for life in all its forms. Animals and their habitats should be cherished, regardless of whether or not they appeared in a Red Data Book (lists of endangered or threatened species issued by IUCN) or could be made to pay their way as tourist spectacles or trophies. Species at risk of extinction, like the panda, tiger, blue whale or elephant, were especially worth fighting for, it was true, but their plight stood for that of all their kind.

Nicholson could get as worked up as anyone about the morals of species survival. In 1971 he raged at the International Whaling Commission for showing 'criminal negligence' by failing to protect the world's whales. Some twenty years later he felt free to admit to Fred Pearce in *Green Warriors* that he, too, was a muesli eater at heart. 'People are turning back to find things in the roots of human evolution. Perhaps the environmental movement's uncanny success has been due to this great wellspring. Politicians are not attuned to this but people in the street talk more sense about the environment than people in Whitehall,' he reflected.

Nicholson and his colleagues also knew that using charismatic figurehead animals like the panda was a surefire way to ensure a steady flow of donations, even if the money raised would actually be spent on projects directed at conserving habitats or creating alternatives to patterns of human behaviour, as well as at protecting 'star' creatures.

Nicholson was unable to square the two tendencies – 'wise use' and 'poetic justice' – within WWF. He couldn't do it because it couldn't be done – as just about every other organisation that followed WWF's path soon found.

Outside the rather circular debate surrounding wildlife and nature conservation issues in the 1960s, broader awareness of environmental ills was being spurred on by events. In Britain, for instance, the public was deeply shocked by the first major oil pollution incident in the British Isles, when in 1967 the tanker *Torrey Canyon* was holed off the Isles of Scilly, causing an oil slick that ruined dozens of resort beaches and stretches of wild coastline, killed thousands of seabirds and blighted seabed and other marine life over a wide area. Royal Navy jets finally bombed the stricken ship in an attempt to burn the oil away. Few who watched this on television could easily forget the image. Was this war? It certainly looked like it.

Another critical occurrence was the tragedy of thalidomide, a drug widely prescribed to control morning sickness in pregnancy, which proved to be responsible for causing shocking deformities in thousands of newborn babies in 1962. Then there were scares over rising levels of the radioactive isotope strontium-90 in milk, linked to contaminants from nuclear weapons testing in the atmosphere. A wave of reports of mass deaths of pigeons, pheasants, foxes and other farm wildlife turned out to be connected to the use of pesticides containing organochlorine compounds, used as seed dressings. Here were the darker flip-sides of the pharmaceutical advances which had delivered the contraceptive pill, the 'green revolution' of agricultural development in the Third World, the *pax atomica* and other boons hailed by Harold Wilson in 1963 as benefits forged in the 'white heat' of technology.

David Brower had a spiel he called The Sermon, in which he first made the comparison – by now perhaps rather overworked but an innovative eye-opener in the late 1960s and early 1970s – of the story of life on Earth as if it were condensed into the six days of the biblical Creation. As John McPhee reports in *Encounters with the Archdruid*, it went like this. On the scale of 666 million years to each one of the six days of Genesis, it took 'all day Monday and until Tuesday noon' for Creation to get the Earth going. Life began at noon on Tuesday and developed steadily over the next four days.

> At 4 p.m. Saturday, the big reptiles came in. Five hours later, when the redwoods appeared, there were no more big reptiles. At three minutes before midnight, man appeared. At one-fourth of a second before midnight, Christ arrived. At one-fortieth of a second before midnight, the Industrial Revolution began. We are surrounded with people who think that what we have been doing for that one-fortieth of a second can go on indefinitely. They are considered normal but they are stark, raving mad.

Brower would then tell his lecture audiences that they and he were hooked on growth.

> We're addicted to it. In my lifetime, man has used more resources than in all previous history. Technology has just begun to happen. They are *mining* water under Arizona. Cotton is subsidised by all that water. Why grow cotton in Arizona? There is no point to this. Why grow to the point of repugnance? Aren't we repugnant enough already?
>
> The United States [he would remind them] has six per cent of the world's population and uses 60 per cent of the world's resources, and one per cent of Americans use 60 per cent of that. When one country gets more than its share, it builds tensions. War is waged over resources [while] we need an economics of peaceful stability.

Referring to the threat of climate change, he'd conclude ironically that yes, there was a human population problem, 'but if we succeed in interrupting the cycle of photosynthesis, we won't have to worry about it. Good breeding can be overdone. How dense can people be?'

These passages convey but little of the flavour of Brower's crusading style, likened by some to that of revivalist preachers like Billy Graham but a good deal wittier. However, the metaphors and arguments Brower used in The Sermon became the philosophy which underpinned the foundation of Friends of the Earth in 1969. And every presentation he made after 1969 would be illustrated by photographs of the Earth from space, brought home in 1968 by the Apollo 8.

Some issues never go away. The issue that finally drove Brower to quit orthodox nature conservation advocacy and set out on the campaign trail under a new banner was one of the most vexed of modern times, nuclear power.

Brower had started out *supporting* nuclear power as a smart alternative to more of the hydro-power dams he had seen flooding his beloved canyons and sierras all over California and in neighbouring states. But by the late 1960s he was challenging orthodox wisdom on nuclear energy, not only by reason of the profound uncertainties surrounding its safety but also because he believed that the massive capital investment it needed could yield far greater net energy gains if spent almost any other way.

Plans had been announced to build a nuclear power station at Diablo Canyon, by the Pacific coast. Brower raised this issue with the Sierra Club Board and called on them to oppose the scheme. The Board declined to get involved. Brower resigned. Within a few months he re-emerged as head of FoE, holding an inaugural press conference in the group's first office, an old firehouse in San Francisco, on 15 September 1969. The new name intrigued the media. The group's agenda was new, too, or rather it was a very new combination of some fairly new and some age-old issues and themes. Nature and landscape preservation were in there, but so were nuclear power, supersonic transport, industrial waste dumping and, fundamentally, an overall consideration of the relationship between humanity and its home planet.

Brower pushed ahead with the first manifesto of FoE, a collection of pieces by various more or less famous writers, thinkers and troublemakers on a range of environmental topics. It was published in the USA by Ballantine Books on Earth Day in April 1970, under the title *The Environmental Handbook*. The range of views presented in the *Handbook* was quite extraordinary. Doomwatch pieces alternated with poetic addresses to nature, scholarly expositions on the science of ecology and, importantly, reasoned alternative courses of action for all. By bringing out the book to coincide with Earth Day, Brower expected to get a mild fillip of publicity for his magpie manifesto from the event, which was orchestrated mainly by the Environmental Defense Fund. Civic groups, schools and other neighbourhood institutions across the USA were challenged to demonstrate their concern for their environment by organising a public event of some sort, no matter how small-scale.

The Earth Day initiative took off beyond the organisers' wildest dreams. Several thousand schools and colleges around the nation put on exhibitions while, at another extreme, a giant 'eco-fair' blocked off large sections of Sixth Avenue in downtown New York. The media went wild and America woke up next day to be told that it had a mass environmental movement. FoE was the biggest beneficiary of this revelation.

Much encouraged and now moderately enriched by book sales, Brower came to Europe in the summer of 1970, to get busy on networking the organisation abroad, starting in France with help from two Americans, a physics student named Amory Lovins and a Paris-based lawyer, Edwin Matthews.

The upshot was the establishment of Les Amis de la Terre in France under a young economist called Brice Lalonde (the same Brice Lalonde who would later become Mitterand's Minister of Environment) and another counterpart organisation in the then Federal Republic of (West) Germany.

In Britain, things were obviously on the move, though not in the specific area of environmental activism so much as student activism and radicalism of an omnidirectional cast. A certain amount of pro-environment signalling was also going on in the fringe and alternative press, in scandal sheets like *Vole*, in the *Ecologist* (a monthly started up by Teddy Goldsmith), and in a conventional consumer-type magazine, *Your Environment*, launched by a group of nature buffs that included the poet Ted Hughes. So there was smoke – but was there fire?

Edwin Matthews was wondering how to approach the task of finding a UK base for FoE when, during a holiday in the west of Ireland, he ran into a retired Scots businessman named Barclay Inglis, who had headed a division of the UK Milk Marketing Board. The pair met on Clare Island, off County Mayo, on a late night trek to watch grey seals. They got chatting after the seals declined to show up. Later, Inglis undertook to help Matthews find committed young activists to form an FoE national chapter in Britain. It took him just three months.

David Brower came to London in the autumn of 1970, determined not to go away until FoE UK had taken at least provisional shape. Inglis invited a group of people to a convivial gathering at the exclusive Travellers' Club in The Strand. After dinner, Brower gave them The Sermon.

Among the fourteen guests were Graham Searle, Vice-President of the National Union of Students, and Jonathan Holliman, a young writer and campaigner then working for the International Youth Federation for Conservation. The next day they contacted another student activist, Richard Sandbrook, former student union president at the University of East Anglia. All three had tried to get the NUS involved in the issues, and failed. Inglis had heard about an eloquent presentation Searle had made, with writing and research support from Holliman, at a 'Countryside in 1970' conference held earlier in the year at London's Guildhall. Catching wind of Brower's visit and the forthcoming session at the Travellers' Club, Holliman had contacted Inglis and asked if he could invite himself along with a couple of like-minded friends. Fine, Inglis had said.

By the end of the evening, Brower had decided there was no need to look further for UK agents and Inglis had picked up a bill for just under £100, 'mostly for liquor', Matthews later recalled. 'Before the evening was over we resolved to start Friends of the Earth UK right then.'

Figure 4 Founder David Brower with the FoE's first Director, Graham Searle, 1970

Richard Sandbrook, who today runs the International Institute for Environment and Development, was a trainee accountant at international financial consultants Arthur Andersen when the call came from Brower and Searle. He was asked to be the equivalent of company secretary, to look after accounts and other aspects of the new group's administration.

'We then used to meet about once a week through that winter and then on into 1971 till we actually became incorporated. We'd get together in the White Lion in Duke Street,' he recalls. 'Various people were involved, a guy called Mike Denny of the *Resurgence* lot, Robert Allen of the *Ecologist*, everybody used to turn up. These were FoE evening sessions, strategy talk.' Sandbrook, whose slight figure and quiet manner are contradicted by a gimlet-eyed attentiveness, has been labelled by friends and enemies alike as something of a modern Machiavelli. His contribution to the fledgling group's early policy formation amounted to a good deal more than figuring out how to keep the books, though that was tricky enough. 'At the Travellers' Brower had said yes, you can use the name and sort of license it and Ed Matthews drew up a licensing agreement, using its California name', he says. 'The first thing Graham had to do was to find income so we were given American books to "translate" into English, like *The Environmental Handbook*, *Frail Ocean* and *The Case Against Concorde*. And Jon Holliman was of course there, he had offices in King Street, shared with Ballantine Books.'

After some persuading, Brower agreed not to insist on a policy structure administered from a US head office. From the beginning, the principles

Figure 5 FoE's first direct action: the bottle dump outside the London offices of Cadbury Schweppes, May 1971

of devolved decision-making and local autonomy have shaped the organisation and structure of FoE. 'The deal was simple, just "railway lines" to live within. There was no money involved,' says Sandbrook. He continues,

> The early campaigns were a hotchpotch of things. Some wanted to concentrate on transport, others on stopping mining in Snowdonia, some consumer recycling, looking after whales, all that. But the whole thing didn't really gel till Graham – bless his heart – stood up at the Institute of Contemporary Arts, where they were holding a public seminar about the environment, and said: 'Well I'm going to take my bottles Saturday morning over to Cadbury Schweppes.' Schweppes had just announced they weren't going to use returnable bottles for their drinks products and people had been vocal about this through the seminar. He said 'Anyone else who wants to do it can come along, FoE is going to organise a bottle dump on Cadbury Schweppes'.
>
> The first time we only had enough to fill twelve handcarts, borrowed from Covent Garden, which we pushed up Park Lane. The second and the third dump were bigger, we took truckloads and the story was out. But the first was enough: we got 50 yards of bottles quite closely set, it looked like a phenomenal sea of bottles. It made a terrific photograph.
>
> It went straight into the Sunday papers and that was that. People started ringing us up in their hundreds. We were away.

5

Poetic justice

The environment is the room, the flat, the house where you live: the factory, the office, the shop where you work: your road, your parish, your village, town or city: Britain, Europe, the world – even the space the world sails through. It's the street where your children play, the park they take the dog in, the flowers, the trees, the animals and birds, the fields, the crops, the streams, the waterfalls. It's the fish, the cliffs, the seashore, the sea itself, the hills and the mountains, the pubs, the bingo halls, the lanes, the motorways, the highways and byways, the farms, the rows of shops and houses, the dustbins, the historic buildings, the trains and buses and cars . . . It's the insects, an empty tin can, aeroplanes, pictures, pollen and the leaves that fall from the trees . . . It's the air you breathe, the blue sky, peace and quiet, the clouds and the sun.

– Barclay Inglis, 'Me save the environment?', in *The Environmental Handbook*, UK version, 1971

A striking oddity that attended the birth of all the more proactive large environmental groups of the 1960s and 1970s was that the defining actions, the campaigns that gave them proof of identity were in many cases wasted efforts. Yet out of these reverses each somehow emerged the moral victor, with a public profile raised beyond normal expectations.

The Schweppes bottle-dump action did more than any other to establish Friends of the Earth as a force in Britain. Yet it forced no change on Schweppes. That didn't seem to matter. In fact, it set the fledgling organisation up as a credible voice in any environmental controversy from then on.

Nothing has baffled historians of the green movement more than the counter-intuitive effect such apparent fiascos have so often had, transforming people's way of seeing the world and, over time, rewriting the standard operating procedures of government, industry, commerce and the law. How come?

To anybody who has read their Marshall McLuhan, the paradox exemplified by the bottle dump's success was no great mystery, just another volatile aspect of daily life in the 'global village' created by the advent of mass electronic and visual communications media.

For the core of waged activists and volunteers staffing FoE's tiny office in King Street, the Schweppes action meant a tidal wave of mail and phone calls from would-be members or volunteers and from people keen to set up local groups around the country. For them, actions like the bottle dump were far more than media-friendly images. They summed up the new values of the environmental ethic, the values they believed should drive the evolution of society.

Some members of the King Street crew had to deal with all this while nursing week-old hangovers sustained in the attempt to build up a critical mass of empty tonic water bottles to arrange on Schweppes' doorstep. Pete Wilkinson, assigned by Graham Searle to collect 2,000 of them, recollects: 'We had about a week to go and simply couldn't find enough of the damn things, or couldn't find enough hours in the day to collect them.'

'We had to go out and buy loads of Schweppes drinks, which we promptly poured into plastic containers. Graham produced a lot of gin and we drank it with tonic in order to justify our purchase. Wouldn't happen today, perish the thought,' he adds, pulling a face to signify the forces of moral correctness.

Wilkinson, who found out about FoE when a flatmate happened to lend him a copy of *The Environmental Handbook*, had drifted in and out of a score of labouring, lorry-driving and dead-end office jobs since leaving school. When he turned up at King Street and Searle offered to sign him on it was, he says,

> like manna from heaven. I think it was always in me, a concern about environmental things but it was latent, I couldn't really identify it. Reading the book brought it all to the surface. And working in Friends of the Earth was vibrant, you didn't know what was going to happen next. You got to do things totally out of the ordinary. I never looked back.

Wilkinson, who later became one of Greenpeace's most effective international campaigners, nevertheless felt somehow stereotyped as the corporation roughneck in the cast of characters at King Street.

> Friends of the Earth was attracting a lot of high-powered types so I always felt typecast as a sort of grammar school educated but basically working class lorry driver. Whenever they listed people that worked for the organisation, they always squeezed me in between Amory Lovins and Walt Patterson to show we had width . . . I always felt I was the arse-end of the organisation.

Lovins was working in England on a scholarship, pursuing postgraduate physics studies at Merton College, Oxford. A grizzled veteran of the two-year-old organisation his friend David Brower had founded, he was also a channel of communication between the US group and the new chapters in Europe. By 1972, these included counterparts in France, Germany and Sweden as well as in the British Isles.

The other North American on the early roll-call of FoE UK was Walt Patterson, a colourful émigré Canadian with a phenomenal, total-recall memory for facts and figures and a passion for pickled gherkins. He also had a gift for stirring up trouble in the corridors of power, a Master's degree in nuclear physics and a wry sense of humour. Patterson has gone on to become a leading authority on nuclear and other energy issues.

Patterson had been in Britain since 1960, having come over initially to complete a PhD in nuclear physics at Edinburgh. But after seven years spent in undergraduate and Master's studies and a year of hanging loose in Greenwich Village, he was in no mood for yet more formal education. Over time, the mainstay of his work at FoE would be energy issues, especially nuclear power. He says he and Amory Lovins took nuclear issues on for some years 'as a sort of "mean cop: nice cop" double act, Amory frightening the wits out of people while I'd go make them a cup of coffee, tell them it wasn't so bad. We worked very well together.'

Patterson had been alerted to the pro-environment buzz by an old friend, Bob Hunter, an important figure in the independent green movement who eight years later was to become founding President of Greenpeace in Canada. Hunter visited Patterson in England, arriving out of the blue in 1968. 'During that visit, he brought home to me what a wave of excitement was building up on the West Coast of North America about environmental issues. I began to get seriously involved,' Patterson recalls.

> In December 1969, I was on the top of a bus going around Hyde Park Corner and I picked up a discarded copy of the midday edition of the *Evening Standard*. It had a little editorial about a new magazine called *Your Environment*, founded by – among others – Ted Hughes and a couple of other poets. I thought, this sounds really interesting. I eventually got in touch with them, sent them some material and offered to write for them on, amongst other things, radioactive waste. The upshot was, I wrote this article for them in summer 1970, my first published piece, called 'Odourless, Tasteless and Dangerous'. Within a year I was the magazine's co-editor.

Eventually Patterson got to be senior editor of *Your Environment* but by then it was owned by a publisher in the East End whose major concerns were soft pornography and catalogues of heavy plant and machinery. The magazine slipped down the publisher's order of priorities and Patterson gave up on it. By that time, however, with the magazine as his calling card, he had begun to orbit in pro-environment circles.

He had made contact with the embryonic FoE, initially through Amory Lovins. He had met Graham Searle, Richard Sandbrook and Jonathan Holliman in the smoke-filled broom cupboard over the Ballantine Books office in King Street where Holliman was still ferreting away at the UK edition of *The Environmental Handbook* with backup from the group's first woman member, Janet Whelan. But Patterson little thought of working with them at this stage. Since 1968, preparations had been underway for a major United Nations Conference on the Human Environment in Stockholm in 1972. The international arena was where the main action seemed to beckon.

Six months after *Your Environment* had started publication, Teddy Goldsmith had launched the *Ecologist*. 'We had a friendly rivalry because we were the only two environmental magazines at the time,' says Patterson. In early 1972 there was a party at Goldsmith's, about the time that the *Ecologist* published *Blueprint for Survival*. At the party Patterson got together with Richard Wilson, a *Times* cartoonist who had also done work for the *Ecologist*, and Robert Allen, then second-in-command on the magazine. They decided they were going to publish an unofficial, independent daily newspaper at the Stockholm Conference. Thus the *Stockholm Eco* (pronounced 'Echo' so it sounded like a newspaper but with ecological knobs on) was born.

The fundamental dilemma to be debated by governments at Stockholm was, briefly, as follows. In the foregoing hundred years, the planet's human population had tripled, the global economy had grown twenty times over, world industry had expanded fiftyfold and the burning of fossil fuels like coal and oil had increased thirtyfold. This runaway industrial growth had triggered a rush towards urbanisation, the crowding of most of the world's people into towns and cities and out of contact with the land and its resources.

For many, changes like these had meant a tremendous hike in living standards and a bonanza of labour-saving conveniences. Yet these same changes, which nobody had ever planned or taken time out to evaluate, were also beginning to have fearsome environmental drawbacks. Activities undertaken in the name of progress now seemed to threaten the very resources on which industry relied, such as the planet's finite metal and fossil-based fuel reserves. Meanwhile, renewable natural resources like forests and fisheries were no longer getting the chance to regenerate as they normally should; the demands of mass consumption were as insatiable as those of mass production.

Runaway industrial growth in the more affluent parts of the world had cast a general blight on the environment, in the tangible form of pollution. Chemicals and wastes from mismanaged farms, factories, urban sewage systems and power plants were fouling air, land, sea and inland waters. Was the 'affluent society' of postwar decades worth so high a price?

At least it might be said to pay off in terms of improved living standards for some in the world's more industrialised regions. In the developing countries of Africa, Asia and Latin America, most people could not look forward to any such dividend. These countries had to contend with an added threat they couldn't buy their way out of: exhaustion of natural resources like forests and fertile soils, heralded by cruel famines and the creeping spread of desert conditions.

These were problems of mismanagement, too, but it was a frailty largely forced on fast-growing numbers of very poor people by land hunger and lack of the purchasing power they needed to fashion secure and rewarding lifestyles for themselves. Of course, widespread corruption and military domination in many countries didn't help. But without economic growth, driven along by profitable returns on capital investment and global trade, how could the world's surplus of poor people find hope of prosperity? Yet unbridled economic growth lay at the heart of environmental problems: how could it also be the solution?

In neither world, Rich nor Poor, were the strains created by this anomaly doing natural habitats, species and ecosystems any good: extinction was stalking the planet. Worst of all, the situation menaced future prospects for preserving the global life-support systems, the oceans and atmosphere, forests and icecaps. On these the survival of human – or any – life ultimately depended. In the end, neglecting them was incompatible with the future of life on Earth, let alone the wealth of nations.

Many saw Rich World–Poor World inequality as the main engine of this destructive slide. Only if there were a more even-handed deal for economic progress could benign development be on the cards. The Poor World actually owned most of the planet's natural resources but the Rich World had a stranglehold on ways to make money out of them. Surely a fair exchange ought to be possible.

The only suitable forum for assessing the world economic situation in terms like these, or to negotiate international measures to combat pollution, was the United Nations. The suggestion had arisen in some quarters that IUCN might step forward to claim this role but at that time its credibility among developing country leaders just wasn't up to scratch.

There were existing UN bodies, specialised agencies like the Food and Agriculture Organization or the World Health Organization which already dealt at least in passing with some key aspects of environmental management; solving development problems was in their nature. The problem was, they were much *too* specialised as they stood to take all the issues on board.

Pressing the case that the UN system courted irrelevance unless it started growing new organs to take cognisance of the new environmental imperatives, the Swedish delegation to the UN had initiated a resolution to convene a major UN Conference on the Human Environment in Stockholm. A poacher-turned-gamekeeper Canadian industrialist, Maurice Strong, was drafted in to run it.

Strong set up an official Secretariat for the purpose and got himself appointed as Executive Director. To help set the tone of the debate which would take place at Stockholm in June 1972, he commissioned an overview work outlining the main problems and opportunities. It was the work of an impressive duo, economist Barbara Ward and biologist René Dubos.

Only One Earth by Ward and Dubos is still probably the most concise and ringing statement ever made of the essential case for seeing environment and development concerns as one. Though it seems dated now in some respects, the main arguments it presses have nearly all been borne out emphatically by subsequent events. It is also a memorably worded book, for Ward was an expressive advocate for the poor and believed passionately that only a better deal for all would stick.

Dubos, though clad in respectable scientific credentials, was also a poet and prophet with a profound belief in 'the importance of developing the distinctive genius of each place, each social group and each person – in other words of cultivating individuality.' He did not see this idea as incompatible with attempts 'to develop the global state of mind which will generate rational loyalty to the planet as a whole.' In fact, insisted Dubos, these endeavours corresponded to two complementary attitudes. As we entered the global phase of human evolution, he declared, it became obvious that 'each of us had two countries, our own and the planet Earth.'

Dubos' heart-on-sleeve scientific humanism was considered by many as an overly sentimental extrapolation from concerns they saw essentially as problems for the plumber. Ward and Dubos realised, however, that a dourly materialistic approach would not win hearts and minds. They also saw an overriding need to honour and cherish cultural diversity, to acknowledge that Western science-based culture was not the only ideal in town.

Thus they set forth UNCHE's main tasks at Stockholm with a much broader remit in mind than international relations had so far embraced. The two foremost tasks, they said, were 'to formulate the problems inherent in the limitations of the Spaceship Earth and to devise patterns of collective behaviour compatible with the continued flowering of civilisations.' They added that the plural 'civilisations' rather than the singular was used deliberately to imply that no one society had anything like all the answers to humankind's dilemmas.

'Just as individual human beings differ in their life and aspirations, so do social groups' and this variety (said Ward and Dubos) was evident in the range of views expressed by the more than 150 consultant experts from fifty-eight countries around the world whose views they had canvassed for inclusion in the book (one of these consultants was, incidentally, Jonathan Holliman). It is worth dwelling on Ward's list of examples of this range of views, for it has not changed a whole lot since.

The leading question was: what were we to make of the effects of technological intervention into the human environment, and what should we do about it? Some respondents said they were more impressed by the stability and resilience of ecosystems than by their fragility and felt no response was called for. Some wanted to emphasise human settlements as the heart of the matter rather than natural ecosystems and nature conservation. Some would prefer to give prior attention to water pollution, others to the state of the atmosphere, still others to land management. 'Some', the list went on, 'see the solution of environmental problems in more scientific knowledge or better technological fixes.' This paraphrased Buckminster Fuller's view that technology ought to 'learn to do more with less'.

'Some believe that environmental pollution and the depletion of natural resources can best be controlled by individual behaviour, others by strict controls over industry and still others by a complete transformation of the political structure or of lifestyles'. All three views would surface in formation in the 1980s, in the political language of the Greens.

> Some believe that the most destructive forms of ecological damage flow from types of high-energy, high-profit technology whose advantages are grossly overstated in terms of genuine utility, others see energy as *the* key to the basic economic achievement of producing more goods for fewer inputs and thus incomparably widening the citizen's wealth and choice . . .

Finally, Ward and Dubos looked at the concepts used to define 'development' and its opposite. Some respondents, they said, disliked the phrase 'developed countries' because they believed no part of the world was yet adequately developed. 'Others, in contrast, believe that industrial development has gone too far in the affluent countries and must be reduced within limits determined by man's ability to stabilize the economy of the earth's resources.'

Ward and Dubos concluded that there was, at least, general agreement amongst all the 'scientific and intellectual leaders' they had consulted, that international action was overdue, that 'environmental problems are becoming increasingly worldwide and therefore demand a global approach' to problem-solving.

But some Third World correspondents thought developing countries should be left to determine solutions by themselves, working within cultural traditions and aspirations that had meaning and value for them rather than at the bidding of outsiders. Finally, Ward and Dubos noted that some sources felt 'the general tone of *Only One Earth* is too pessimistic'. The whistle-blowing approach typified by Carson's *Silent Spring* should, various consultants within industry warned, be discouraged as 'emotional and non-factual' scaremongering.

Some correspondents, however, urged that the book should be *more* forceful in denouncing adverse environmental trends they felt were propelling mankind towards self-destruction. This cluster of views was amply represented in *Blueprint for Survival*, which came out just before *Only One Earth* and to a certain extent upstaged it. For the effort Ward and Dubos made to cater to all camps hampered them from offering drastic prescriptions for change.

No such inhibition stayed the hand of the authors of the *Blueprint* (Robert Allen, Michael Allaby, Peter Bunyard and Teddy Goldsmith), which would be hailed by a later Director of FoE, Tom Burke, as 'the seminal document for the birth of the modern environment movement. It created a framework of ideas for the first time. The *Ecologist* lot, a very important group of thinkers, put it together, making it up as they went along.'

Not only did the *Blueprint* state the problem in a nutshell, denouncing industrial growth point-blank as the enemy of environmental care, it also proposed remedies. It made a plea for recycling more waste products and offered ideas for replacing mass production and consumption with more compact and accountable systems of resource management and exchange, a view Fritz Schumacher's book *Small is Beautiful* would refine a year or so later. In the Stockholm context, the *Blueprint* was a timely and noisy wake-up call. Its galvanising impact owed much to its format, a compact magazine supplement, tersely journalistic and visually alive. Had the effective unit of comment begun to shrink from book-sized to bite-sized?

That thought may have partly influenced the design of the *Stockholm Eco*, though Walt Patterson says the planning side of the newspaper was fairly haphazard. He recalls arriving at Stockholm to find FoE more or less running the non-governmental lobby by way of the *Eco*, which was distributed free to the hotels of all the delegations. Leading the effort was a team from FoE's San Francisco office that had recently launched a monthly newspaper called *Not Man Apart*.

Their Swedish and British counterparts were also much in evidence in the NGO Forum and on the *Eco*. Graham Searle wrote a wickedly funny gossip column called 'Gremlin'. The Swedes provided an office and 470 bottles of beer. It was a hot June and most nights the work went on till three in the morning without a break. This way of working became a tradition for the FoE teams that would continue to produce *Eco* as a lobbying tool and daily record of fact at the decision-making intergovernmental meetings throughout the 1970s and 1980s.

The Stockholm experience was, Patterson says, 'one of the most electrifying episodes of my life'. What the *Eco* crew pulled off, with the tacit support of many of the Third World delegations to UNCHE, was to virtually hijack the agenda. The conference was fixed well ahead of time. Strong had set up very elaborate Secretariat arrangements. For the first week of the conference nobody paid much attention to the *Eco*. The wild card was the arrival of the Chinese.

> It was the first conference they'd attended since taking their seat in the Security Council. They immediately announced they didn't like the draft text of the conference Declaration. This caused consternation. A parallel closed session had to be set up to redraft the text. We had a mole who leaked us the Chinese text while the session was still running. That first Friday of the conference we ran it in the *Eco* under a banner headline 'CHINA DECLARES' and sat back waiting for the shit to hit the fan.
>
> Stockholm was swarming with media from all over the world. But Saturday came – and Sunday went. Not a nibble. On Monday the story came out officially – and we'd had it forty-eight hours earlier. From then

> on every morning we were besieged by journalists from every paper in the world, all the big TV networks. We had a daily editorial meeting in the Delegates' Lounge and we were four deep in journalists every morning.
>
> They were asking: 'Where's the action? What's happening today, who's doing what to whom?' From then on, for the rest of the conference, the *Eco* was required reading. It really was exciting stuff.

But did it substantially change the agenda? Patterson feels sure it did. 'We were reporting on where the controversies were, who was trying to pull a fast one, what the agenda was, what the hidden agendas were,' he explains.

Sympathetic press coverage multiplied this low-down and levered influential decision makers and power brokers at Stockholm into more radical positions than they'd bargained for. The science editor of the London *Times*, Pearce Wright, led his report on Stockholm with the headline 'FoE EXERTS CRUCIAL PRESSURE'.

The bottom line at Stockholm was not, in fact, a specially dramatic bid for a better world there and then. Structural concerns headed the agenda. A global institution was needed to fend off environmental nemesis in a coordinated manner. This also meant taking steps to set Rich and Poor on level terms in a common cause, environmental security for a small planet.

The idea of a separate, specialised UN agency dealing with the environmental aspects of development was rejected as a possible means to this end. Nothing less than a worldwide and system-wide coordinating body would do, with responsibility for monitoring the environmental impact of the development deeds of all UN agencies and member states.

Parties to the Stockholm Conference duly approved the setting-up of this body – the UN Environment Programme – subject to ratification by national legislatures, a hurdle which would take some time to clear. UNEP was finally set up in 1974, in Nairobi, Kenya. This location was chosen to reassure developing country interests that the new body's functions would not be subverted by 'environmental imperialism' in the interest of former colonial powers, or of the dominant East or West military blocs, or of big business.

Many developing country leaders had come to Stockholm with grave suspicions that the pro-environment agenda might be an undercover plot against their development. One delegation listened to complaints about the polluting industries of North America and northern Europe, then asked if they could help by taking those industries home with them: they needed the jobs.

In July 1972, when everybody had returned from Stockholm, Graham Searle invited Walt Patterson to join the regular staff of FoE. In those days the salary was the unprincely sum of £1500 a year across the board, regardless of status or length of service. Patterson's wife, a dentist, was heard to remark at a staff party later in the decade: 'Well, a lot of people give money to Friends of the Earth: I gave Walt to Friends of the Earth!'

Most of the group's original staffers survived financially either (like Patterson) thanks to the support of a spouse or by doing two jobs back-to-back. Richard Sandbrook was still serving his articles in accountancy with Arthur Andersen, for instance, while Holliman was still employed part-time by IYFC. Among Sandbrook's souvenirs are several pay cheques from those early days. Knowing the financial position of the group only too well, he never banked them.

Along with Searle, Wilkinson, Whelan and Patterson, the shoestring King Street payroll soon included wildlife campaigner Angela King, lawyer Oliver Thorold, and a press officer, Colin Blythe. Others, like wildlife expert John Burton, got involved in particular campaigns, then moved on. The core staff was just seven strong, the 'Magnificent Seven' referred to in an article by *Guardian* columnist Jill Tweedie, one of many good friends of the Friends in the media of the day.

Other media backers were Pearce Wright on the *Times*, Anthony (Phil) Tucker of the *Guardian*, Jeremy Bugler of the *New Scientist*, the *Observer*'s Gerald Leach, and Kenneth Allsop, a TV journalist prominent on BBC prime-time current affairs magazine slots, who had written a hard-hitting Foreword to *The Environmental Handbook*. So although FoE was a minuscule outfit, it had some magnificent multipliers.

An honorary Eighth to the King Street Seven was Barclay Inglis, who continued to serve FoE UK in a kind of Dutch uncle role, then, when a formal Board of Directors was established, as its unflappable Chairman.

Tom Burke, later a Director of FoE, feels Inglis played a crucial part in establishing the group's early identity.

> I think it was Barclay who established the key concepts behind the way Friends of the Earth was run: the focus on research, the concentration of resources, the absence of a membership in a sense normally used to refer to members of a society. Barclay was responsible for a lot of these things and for guiding Friends of the Earth – very deftly – off the ground.

But the group's dynamo and indisputably its leading light was Graham Searle. 'One of the real founders of the green movement, a driving force,' says Pete Wilkinson. He remembers Searle as 'a very hard taskmaster who didn't brook foolishness or incompetence. Deep, too: he could argue about football one minute on the street corner with the market porters, next minute he'd be on TV slamming Rio Tinto Zinc. An absolute past master at debate, a very powerful force.' He needed to be.

> It was breaking new ground time, you didn't have any points of reference. It put a lot of strain on everybody. Graham was there steering it right the way through. It needed that driving force behind it, that strength of

Figure 6 Administrator Angela Potter (seated at back) and colleagues display the first FoE campaigning materials © McKenzie/FoE

WHO SAYS THE ENGLISH ARE KIND TO ANIMALS?
Madam, your new tiger skin coat is almost ready.
Down with National Parks.
KEEP OFF THE GLASS.
THE SOFT DRINK PEOPLE.
Schweppes
KEEP OFF THE GLASS.

> personality. I don't think Friends of the Earth would actually have got off the ground if he hadn't been there behind everything. A competent guy, a friend.

Strong personalities usually have their downside and Searle's was no exception. According to Richard Sandbrook: 'Graham on one level could really be a shit of a man . . . another side of him lovable and considerate; a Jekyll-and-Hyde character.' Even when Hyde was uppermost, Sandbrook stood in awe of Searle's writing skills. 'He was a very fast, natural copy writer,' he says. 'First of all, though, he was a brave man, not frightened of anybody. Also very shrewd.'

Trained as a geologist, Searle was a stickler for sound science as the best guide of policy. A key motto of his was: 'Get it right. Nobody can hit us if we get our facts right.' 'That's why we always had such strength in research,' says Richard Sandbrook. 'I can remember that as one of the main unwritten rules of Friends of the Earth from the start.' There were, he adds, others which everyone acknowledged.

> One was Always Decentralise – nowadays I suppose you'd call it subsidiarity. Another was that everything had to have humour and humanity in it. You had to get the facts right first, mind, but then you had to have fun. Another principle was that we would work within the law – just. A fourth ground rule that was invented later to break deadlocks was that nobody should be Director for more than three years.
>
> The three-year rule reflected the idea that we wanted to be in a permanent state of revolution, regeneration, change. The organisation shouldn't be allowed to ossify. All very sixties in outlook and long gone.

Sandbrook omits to mention a more general guiding principle behind FoE that has arguably been one of the organisation's main claims to fame and cultural influence down the years: the Think Globally, Act Locally formula. Its attribution is vague, but even if they didn't invent it, early figures in FoE used and popularised it ad lib. It soon came to be associated more with them than with any other pro-environment grouping. But its interpretation was to prove a slippery affair.

6

From the heart

> All of us who started journeying with Friends of the Earth started out of passion, not out of professional careerism, though some of us maintained professions in parallel. What happened all through the 1970s – and we watched it happen – was that the arguments and ways of arguing strayed ever closer to other people's frame of reference. We started outside the frame, saying: Come on, you ought to come over here. But the arguments got to be more and more an exchange of expert opinions.
>
> – Sue Clifford, Friends of the Earth Board Member 1971–82

The Think Globally, Act Locally motto rhymed closely with Dubos' quietly poetic idea that everyone should care for the environment out of loyalty to two countries, their own and the planet Earth. It was to be a cornerstone of policy making and campaigning practice in Friends of the Earth, as well as apparently underpinning the organisation's structure.

That structure's evolution was, however, also decided by more mundane considerations. Seeking charitable status, then the beaten track for voluntary organisations in the UK, was ruled out from the start by the group's founders, for charities were barred from engaging in politics in public. Limited company status was opted for instead, though plans were made to set up research and fund-raising charity affiliates in due course, plans which would bear fruit in 1976 and 1980 with the genesis of Earth Resources Research and Environmental Research and Information Ltd, later renamed Friends of the Earth Trust.

A growing band of local FoE groups also came swiftly into evidence, numbering more than seventy by the middle of 1973. Their structures and functions were left largely up to them once they had been vetted and approved as franchised users of the Friends of the Earth name.

After a hefty dose of Thinking Globally at Stockholm, the UK national chapter of FoE was more than ready to Act Locally by taking on new challenges at home, having established a campaigning style that seemed to work more persuasively than anyone had dared hope. But it also had to establish an accommodation between the high-gloss, high-profile media theatre of the London crew and the nitty-gritty of local campaigning by the nascent local groups.

Another problem the organisation had yet to resolve was the question of whose language it spoke. Sooner or later it had to decide if business, industry and government were to be shunned as out-and-out Evil Empires or engaged in dialogue. Richard Sandbrook neatly defines one way out of this dilemma: 'the polyglot institution' that can speak the language of business, or of alternative lifestyles, or of political change, depending on whose response it most needs to lever at the time. But local action could still, in FoE's case, be the 'mother tongue'. 'Friends of the Earth was – and is – in a powerful position,' Sandbrook asserts, 'to mobilise community priorities, to make them provincial, then national, then regional, then international concerns, to drive for policy from the bottom up.'

The first demonstrably effective action of the national FoE (though success would take a while coming) was a campaign after David Brower's own heart: keeping a mining corporation out of the mountains. In 1972, the transnational mining conglomerate Rio Tinto Zinc was seeking licences to dig a gigantic hole to extract copper from one of Britain's most scenic and unspoilt wilderness areas, Snowdonia in North Wales. The area had been designated a National Park but the same 'mining exception' loopholes Brower had raged against in the High Sierras of America also riddled the mineral rights position in Snowdonia. Searle assigned Amory Lovins and an Australian geologist, Simon Millar, to prepare a measured case against the RTZ proposals then fight them tooth and nail. They were ably assisted by a young barrister volunteer David Bunt and a bright lawyer, Oliver Thorold.

More conventional pro-nature groups, led by the Conservation Society and the Council for the Preservation of Rural England, also raised clamorous opposition to the scheme. The stir led Rio Tinto Zinc to set up a Commission on Mining and the Environment, under Lord Zuckerman. Lovins and Millar submitted a treatise called *Rock Bottom* to it, also published in full in the *Ecologist*. A wave of media criticism crested with a BBC TV *Horizon* documentary called *Do You Dig National Parks?* FoE gave background data to the programme's researchers and the prime-time transmission in May 1972 was followed by a live studio debate in which two RTZ moguls faced Searle and Lovins, whose searing cross-examination left them visibly fuming.

The final report of Zuckerman's Commission fudged and fumbled the RTZ issue. But the whole debate was further delayed in 1973 by referral to the Government-appointed Stevens Committee on National Parks, which FoE also lobbied, supplying detailed arguments against mining incursions into Snowdonia. Allen and Unwin had published a new picture book by Amory Lovins and Philip Evans, *Eryri, the Mountains of Longing* after a court battle in which RTZ sought to halt its publication on the grounds that it contained comments prejudicial to the corporation's case for mining in Snowdonia. This high-handed tactic backfired: RTZ's plans were ultimately stalled by a howl of public outrage, fuelled largely by FoE local groups, who raised the issue's profile nationwide. And yet, though Snowdonia now appeared safe enough, nothing had really altered on the statute books.

To this day, nothing in UK law prevents RTZ or anybody else from turning National Park land into a mining eyesore. If there was a policy lesson to be learned from the episode it was that site-specific protests in the Dave Brower stamp did not necessarily work a cleansing effect on the system under British circumstances. Whenever FoE took on such battles in the future, it would be as part of much more free-ranging strategic campaigns, rather than courting collisions over one site.

Running concurrently with the RTZ tussle was the emerging campaign against excessive packaging. Though it, too, proved inconclusive in a strict sense, it exemplified a style of campaigning that established the British group's home-grown personality. There was the humour and humanity Sandbrook spoke of, which for Walt Patterson and others was a clinching argument for joining up.

'The thing I liked most about FoE was that for the first time here was a campaigning organisation that knew how to use humour. That went for the Schweppes bottle dump, a bizarre thing to do, it just threw everybody,' Patterson reminisces. He contrasts the Schweppes demo, 'a cheerful two fingers raised in the face of big business', to the awesome hostility of anti-Vietnam demos of recent memory. In the Grosvenor Square riots of 1968, for instance, more than 100,000 marchers had fought pitched battles with the police.

The cultural context for this turmoil had been, to borrow Ian MacDonald's expression, 'a brutal *Zeitgeist* shift from love and peace to politics and struggle' in the latter half of the 1960s. What FoE had demonstrated was that there didn't have to be an either/or choice between radical leftist streetfighting tactics or 'tune in, turn on, drop out' escapism. Well-informed opportunism with a dash of invention and daring could be just as potent a recipe for change as any overdose of violent confrontation.

Another key aspect of the Schweppes controversy and the wider war on waste it spearheaded was that it marked the baptism of yet another characteristic feature of FoE, the idea that protest and dissent could go hand in hand with being *for* as well as against things. This was an approach that gave the local group network and the wider public a warmly encouraging cue to get involved in practical moves to apply alternative thinking, to demonstrate that it *could* be done. And it has remained a characteristic of FoE at all levels – national, local and international – as well as across all its campaign areas.

For Pete Wilkinson, who felt strongly that actions spoke louder than words, this made sense of the more cerebral and strategic side of campaigning projects. Wilkinson did his share of wordsmithing, contributing a regular column to the *Ecologist* and writing several formative campaign manuals. But there were, he felt, strict limits on what the written word could achieve:

> You can write and write, but eventually you've got to get out there and demonstrate by example. So I was always on the action side of things saying, you've made your case but now what do you do? Where's the action? Have you translated that into things that people can do? What are the political imperatives? That was how it worked in the early days, thinking laterally, cutting-edge campaigning, confronting authority over their inactivity.

The first thing Patterson did on joining FoE was to rewrite a draft campaign manual on packaging, which Searle, who had a flair for coming up with catchy titles, called *Packaging in Britain: A Policy for Containment*. It was a best-seller, winning acres of media coverage and the keen attention of influential names in politics, business and advertising. 'In effect I became the group's garbage expert because of it,' says Patterson. 'It had lines like: "are we going to come to the day when you can buy a single green pea, shrink-wrapped?" In those days it was a way to bring an environmental issue right onto your kitchen table.' But it didn't end there.

An advertising agency donated poster designs and half-page spaces in the national press to run them. They included the cheeky 'DON'T LET THEM SCHH. . . ON BRITAIN' image, an outline of Britain crammed full of non-returnable bottles, a sly dig at the 'SCHH. . . YOU KNOW WHO' advertisement Schweppes had long traded on. 'Because we had a high media profile,' explains Patterson, 'ad people liked to do things for us. It meant they could do something for a good cause and at the same time demonstrate their skills.'

The advertising campaigns, which became industry award-winners, did not just raise the group's national profile for its own sake: it also lent the power of national 'branding' to the elbow of local groups, many of which took the Schweppes campaign to the provinces, picketing Schweppes' distribution warehouses around the country. All the groups participated in a series of National Packaging Days and some went further on their own initiative, setting up demonstration recycling centres and urging good packaging practice on local retailers. Data from *A Policy for Containment* were on hand to back their message up.

'Although we started out simply to pillory Cadbury-Schweppes, we eventually realised that this issue was not just a question of returnable bottles, it went right back through the whole system', comments Patterson. Maybe the campaign action had missed its initial mark but the point behind it would hit home over time in hundreds of experiments in local action against waste, giving all involved the thrill of breaking new ground.

Breaking ground in a more literal sense was the focus of another 'lifestyle' campaign of the 1970s which grew, thanks to Pete Wilkinson's zeal for action, from words which might otherwise have lain fallow. Colin Blythe had written a paper with Michael Allaby and Chris Wardle called

Losing Ground. It hinged on fears then current of an impending world food crisis, pointing up the contrast with wastage in food production systems. Of course, it was a worthy tome as usual', Wilkinson says, 'but I said well, now we've got to translate this into something people can do out there on the streets and get involved in. I badgered and badgered and Colin eventually agreed that we'd look at the situation in London.'

Through a contact in the Greater London Council, Blythe and Wilkinson obtained figures about spare development sites and blighted land around the capital, land left empty following slum clearances in the 1950s, some of it undeveloped since the Blitz. It turned out there was something like 17,000 acres of land that fitted this description. Wilkinson also learned that around 25,000 Londoners were on waiting lists for allotment gardens. He proposed a campaign for turning unused land over to allotments, under an agreement with the GLC that the land would be vacated when they wanted to develop it. 'In view of the food crisis Colin's report had highlighted, we suggested everyone should use every spare bit of space we'd got to produce food, locally-grown produce,' Wilkinson explains, 'and it would also get city folk back to the land.'

Five demonstration allotment spaces were subsequently opened at The Cut, near Waterloo Station, and the campaign then grew to take in other parts of London. Local groups elsewhere around the country picked up on the idea and by the end of the decade thousands of sites had been established all over Britain. Some are still under cultivation to this day. Low-tech, grassroots campaigning like this may not have made the biggest headlines of FoE's early years but it bore out the Think Globally, Act Locally ethic for all to see and many a local group got high mileage from it.

At the other end of the technology spectrum, FoE was spoiling for a fight over the half-developed Anglo-French supersonic transport project, Concorde. FoE US had scored its first conspicuous success in the USA with a national campaign to block plans for an American SST. Could a similar drive work over here?

The Case Against Concorde, by Richard Wiggs, edited by Graham Searle, was one of the first FoE publications to come out in the UK after the *Handbook*. The aircraft might represent the technologically optimum answer to moving people through the air at speeds faster than sound, but who on earth had put the question in the first place? it asked. The cost was soaring. What were the benefits, apart from national prestige?

Certainly nobody had taken the trouble to find out whether the air transport market wanted such a luxury or could defray the vast investment costs with ticket sales. Nor had anybody consulted the needs of householders living near large airports, who would have to put up with aircraft noise at nearly twice the going level yet get nothing back in real amenities they were likely to be able to afford or enjoy.

The book won wide support, not least in the mainstream science community, which was deeply divided over the issue. But there was too much going for Concorde politically to allow the nit-picking of a few boffins to divert the programme from its course, still less the outpourings of a handful of wild and woolly dissidents.

The plane had started out as the very embodiment and flame of Wilson's 'white heat of technology' vision. Its history was also rooted in the siege mentality of postwar science. No less a figure than Barnes-Wallis, of Dambusters fame, had launched the notion that to maintain cohesion and security, Britain needed the capacity to reach every part of its far-flung Empire within a day. Conventional aircraft could not deliver this. (By the same reasoning Barnes-Wallis helped put the development of nuclear submarines on the postwar research and development agenda, though he saw their role principally as merchant vessels that could dodge the enemy in the event of another war.) Provisioning and guarding the British Empire had little meaning now, but the juggernaut was in motion.

By the early 1970s, although the stripe of politics had changed, Concorde had yet another political issue riding on it: Britain's membership of what was then called the Common Market. Following rebuffs from Charles de Gaulle, who had repeatedly vetoed Britain's membership, for Edward Heath to drop the joint Concorde project now would appear to confirm de Gaulle's distrust of *perfide Albion*. None of these factors operated in the USA: many in the US aircraft industry were, moreover, up in arms against SSTs. Here, however, an anti-Concorde campaign had a snowball's chance in hell. Searle and company soon saw this and pragmatically let the issue drop out of sight, though the group won credit in the popular science press and elsewhere for a brave try against ludicrous odds.

One lesson the Concorde experience had taught FoE – at least for the time being – was that its likeliest avenue to success lay not so much in loftily analytical contributions to a high-powered debate over the uses and costs of technology, but in more populist and eye-catching grassroots campaigning 'from the heart', albeit still with adept fact-finding, good PR and sound science at the rear.

In 1971, a few months before the Schweppes action hit the headlines in Britain, a ragamuffin group of art school students and their tutor, Joseph Beuys, were demonstrating against the destruction of an ancient woodland, the Grafenburger Wald, on public ground in the suburbs of Dusseldorf, in the heart of what was then West Germany. The demonstrators wielded witches' brooms which they used to sweep the boundaries of the woodland, to symbolise their repossession of a natural common that was about to be swept surreptitiously aside. The Wald was threatened by impending conversion to new tennis courts for a fashionable sports club much frequented by local politicians.

This art action, ironically entitled *Overcome Party Dictatorship Now!*, attracted avid press notice and effectively stalled the development plan, though part of the woodland was lost to initial clearing operations. The media stir this local action created went national. The action's *aim* was hardly revolutionary: what was new was the surprise use of performance spectacle to clinch its impact and its appeal over the head of the middle class power structure to a newly enfranchised, environmentally aware youth culture through the mass media. No antidote was known to this media theatre or 'Agit-Pop' form of protest, which sprang from an emerging vogue for alternative art 'events and happenings'.

Beuys also undertook a similar defence of threatened Danish wetlands in his *Bog Action* of the same year, justifying it in language which would not seem out of place in today's green campaign literature:

> Bogs are the liveliest element in the European landscape, not just from the point of view of flora, fauna, birds and animals, but as storehouses of life, mystery and chemical change, and preserves of ancient history. They are essential to the whole ecosystem for water regulation, humidity, groundwater and climate in general.

Beuys's employers at the Art Academy of Dusseldorf took a dim view, however, of his advocacy for environmental care and for art as activism. They saw it as trespassing outside the proper territory of art and culture. After several warnings they finally sacked the artist from his teaching post. Their reaction didn't spring entirely from art traditionalism. During the Hitler era of painfully recent memory, artists had been commandeered by the Nazi regime as mouthpieces of the jingoistic 'Blood and Soil' ideology that had been used to justify one unspeakable atrocity after another.

For obvious reason, the politics of Right and Left were kept in strict quarantine during the early postwar years in Germany, a country torn in two first by the aftermath of World War II fascism, then by communism's Iron curtain, a literal presence in Germany in the grotesque form of the Berlin Wall. Yet Germany remained the acknowledged target 'theatre' of any nuclear war, a threat that at times looked only too imminent.

West Germany's landscape and cities, already disfigured by total warfare, were now at the mercy of the get-rich-quick materialism and *laisser-faire* of the reconstruction era, unopposed by organised labour movements. By caving in to no-strike deals the unions had left their populist political heritage lying vacant. It was into this vacuum that Green politics would shortly step, though with an ideology that purported to shun conventional ideas of Right or Left.

A binge of unfettered re-industrialisation and redevelopment had begun to give rise to chronic pollution ills and a related 'visual pollution' problem and social dilemma: the gaunt uniformity of postwar urban, agricultural and highway projects under which rural and small-town Germany were fast vanishing. A disaffected postwar generation felt increasingly trapped by this situation through no real fault of its own.

Beuys strongly sensed that the pacifist concern younger generations felt over the future took precedence over his own generation's backward-looking paranoia and the traumatised antipathy most intellectuals felt towards participating in politics. More than that, as a knowledgeable observer of the life sciences, he felt that engaging in environmental issues was dramatically validated by new scientific ideas of Nature, especially by the ecological world view then being promoted up the hierarchy.

Beuys's answer to his dismissal from the Academy was to continue to offer tutorials on the pavement outside. Later he expanded this enterprise into what he called the Free International University, a 'virtual academy' co-founded with a group of artists and intellectuals that included the novelist and Nobel laureate Heinrich Böll. The FIU's agenda later fused with that of like-minded workers' groupings such as the Organisation for Direct Democracy, as well as comeback socialist intellectual groups and civic amenity or heritage groups of a distinctly conservative cast.

A charismatic firebrand of a campaigner, Petra Kelly, later came to the fore of this chaotic 'anti-party' coalition. As a foreign student at college in Washington DC in the late 1960s, Kelly had taken part in marches for the civil rights and black emancipation movement led by Martin Luther King, and admired the non-violent civil disobedience tactics King championed. She had also been much involved in the US anti-war movement and was also influenced by the fast-growing feminist movement there.

A member of Germany's version of the Baby Boom generation, Petra Kelly was acutely in tune with the disaffection and rebellion astir among her contemporaries back home. She and Beuys formed a strong, albeit unlikely-looking bond. The artist's past was decidedly different from her own; he had piloted dive-bombers for the Luftwaffe in the war, winning the Iron Cross after a string of near-death combat experiences.

In a manifesto for the FIU in 1972, Beuys and Böll had declared:

> Environmental pollution advances in parallel with a pollution of the world within us. Hope is denounced as illusory, and discarded hope breeds violence ... The Nazi Blood and Soil doctrine, which ravaged land and spilled blood, has disturbed our relation to tradition ... Now, however, it is no longer regarded as romantic but eminently realistic to fight for every tree, every plot of undeveloped land, every stream as yet unpoisoned, every old town centre, and against every thoughtless redevelopment scheme. And it is no longer considered romantic to speak of Nature ...

Long before, in 1958, Beuys had scorned 'the silence of intellectuals and artists' who had condoned the horrors of Nazism. Unless artists spoke out as impartial champions of nature and culture, he felt it could happen again. This time round, the enemy was the technocrats, the developers, the NATO warlords and their political stooges who were using their power monopoly to enforce a stifling and exploitative status quo just as life-threatening as Nazism.

Messages like these could fire up traditionalists on the natural Right as much as up-and-coming radicals on what historians might call the rump of the Left. In Belgium and a few other parts of northern Europe where counterpart movements later emerged, Green politics initially took a noticeable Right turn. In Germany, though its leaders proclaimed it a party 'neither of the Left nor the right, but of the Way forward', the Green Party in the 1980s would find the bedrock of its electoral support on the old Left, among skilled workers, younger liberal professionals and politically sensitised students.

Green politics offered a new deal that would call a halt to further blight on the physical heritage of town and country. It owed nothing to the terrorist fanaticism of Baader–Meinhof, nor the staid, guilt-ridden piety of the centre-Right Christian Democrats. It also showed a way of building lifestyles on a socially and environmentally responsible base. That bore recognisable similarities to the cross-currents of argument that were making FoE and similar groups a force in Britain, the USA and elsewhere outside Germany.

But Germany's greening owed little to these groups. Here, science-based idealism re-validated patriotism and heritage. In hindsight, maybe it also nourished an inward longing for Germany's reunification, considering the emphasis ecology places on natural systems that transcend human boundaries. Yet it was far from inward-looking. It gave postwar Germany access to ethical and geopolitical debates, including Third world issues, in which it could join on level terms with the rest of the global community after decades spent in disgrace from it.

The birthing of groups like FoE UK was enacted on a goldfish-bowl scale, then with a bit of luck and imagination magnified by media curiosity. In Germany conditions existed for a deep groundswell of political activity and change. The vacuum on the missing Left of politics was part of these conditions, as was a providential voting system, proportional representation. But the deciding factor would be the decision made by the USA and NATO to site Pershing tactical nuclear missiles on West German soil in the early 1980s. After a decade of false starts and largely self-imposed changes of course, the Greens would triumph in 1983, gaining a casting vote in the Bundestag. For now, the dream was enough.

Meanwhile, back in the Britain of the middle 1970s, the hippy sun was setting. The anarchic and egoistic rebellion of the punk era would soon rise in its place. The Tolkien-inspired hippy eaterie and music club, Middle Earth, just opposite the FoE office, began to acquire a jaded look.

Yet the essential spirit of the 1960s, a hunger for peace-loving and inspirational alternatives to the orthodox and predictable, had found a course of action that gave it somewhere to go and renewable value and meaning. By hitching its wagon to ecology, arguably one of the most nit-picking of all scientific doctrines, it had found an authentic mission in life.

A new catch-all term to describe that mission had also begun to emerge from the undergrowth, though it still wasn't in general circulation: green. It would be a mixed blessing at odd times in the future but as a simple label for a complex project, it was initially a godsend. As to its definition, even those most involved in green affairs couldn't come up with one without fear of contradiction, though that didn't stop the media trying to define their job for them.

In media and public perception, green would become inseparably linked for years to come with images of a huge marine animal. Hardly anyone had ever seen it in the flesh but most people thought of it as blue. The time had come to Save the Whale.

7

More beautiful than death

> Can he who has only discovered the value of whalebone and whale oil be said to have discovered the true uses of the whale? Can he who slays the elephant for his ivory be said to have seen the elephant? No, these are petty and accidental uses. Just as if a stronger race were to kill us in order to make buttons and flageolets of our bones, and then prate of the usefulness of man. Every creature is better alive than dead, both men and moose and pine trees, as life is more beautiful than death.
>
> – Henry David Thoreau, *Journal*, 1853

Angela King thinks it was Christine Stevens of the Animal Welfare Institute in the USA who kickstarted the Save the Whale campaign in the UK. Various groups, including Friends of the Earth US, World Wildlife Fund and the Fauna and Flora Preservation Society, had been working in a low-key way on whale issues for some years before the Save the Whale message became linked in 1970s consciousness to FoE in its British format. What made the connection stick was the inventive way the group handled the campaign – and its wider work on endangered species – as street theatre, pop parable, media war and power game in one. King says,

> I can remember the thinking behind Save the Whale and how it involved symbolism, how the whale symbolised the world. Most of us hadn't seen a whale and would never have much hope of seeing one. But there was something about this enormous creature that symbolised a whole series of things about the world and our relationship with it.
>
> Since then we've all realised how important it is to capture the imagination if we're to keep the interest of supporters and campaigners. We didn't analyse it in those terms then, though, or think through what the campaign would do for us. We just did it. Intuition and passion drove us.

Angela King was working in New York as a designer in the rag trade, campaigning against the use of animal fur and as part of Trash for Cash in Haarlem. But the first Earth Day shenanigans in Manhattan in 1970 really excited a wider interest. She was doing voluntary work for the

Figure 7 Endangered species campaign materials

Like the coat?
The last owner
was killed in it.

Environmental Action Coalition, when somebody visiting from England said, did she know FoE were starting up in London?

When she got back to England in 1971 she turned up at King Street. As she recollects the moment: 'I knocked at the door and there was Graham Searle. I started explaining what I had been doing and he said, did I want a job? I said, yes. He asked what I wanted to do. I said wildlife campaigns. He said, go ahead.'

World Wildlife Fund was the biggest noise in international wildlife issues at the time. WWF was developing a drive to stop the trade in fashion furs, also the topic of King's first campaign. 'But they had a very different way of working,' she explains, 'not a populist approach at all, mostly they were using official networks such as the infant Washington Convention on International Trade in Endangered Species of Fauna and Flora, which WWF had helped set on track, mainly under IUCN auspices.'

Richly patterned big cat furs like leopard and cheetah had become newly fashionable during the mid-1960s, as a delayed reaction against the 'austerity binge' of the 1950s and more recently out of desire to emulate the Jackie Kennedy look. In collaboration with John Burton, King devised a campaign that would take the issue onto the High Street and into the face of affluent consumers who were bankrolling the slaughter of thousands of rare Asian and African wild carnivores month on month. If consumers couldn't be shamed into finding a less destructive status symbol than a flayed cat, there might not be time for international lawgiving to stop the slide towards what looked like almost certain extinction.

A magazine called *Animals* (now *BBC Wildlife Magazine*) lent FoE a mailing list of readers who had responded to an appeal for pledges of commitment not to buy or use wild animal skins. By contacting these sympathisers, the group raised a subscription fund to fight the campaign. That campaign's ultimate goal would be to exert pressure on the Government, by way of public opinion and media pressure, to introduce an Endangered Species Act comparable to legislation on this score already in force in the USA.

Aided by lawyer David Pedley, a supporter skilled in drafting legislation and one of a series of lawyers to join the organisation's payroll, FoE was even in a position to provide the lawgivers with detailed prompts and draft texts for statutory measures. But the real powerhouse of the campaign was growing pressure in the media, then, increasingly, from the public acting as concerned consumers and voters. Angela King recalls: 'We began to realise what we could do with the press. Sometimes we would drink with journalists in the pub and formed friendly relations that way. Or they often just dropped in on us. Top photographers and feature writers arrived on the doorstep and said: What can we do?'

An early media blast on behalf of endangered wildlife had been a *Daily Mirror* front-page report in November 1971 banner-headlined 'OUR GREEN AND POISONED LAND'. The work of a team of *Mirror* journalists advised by Pete Wilkinson and led by *Mirror* man Bryn Jones (who would later run Greenpeace UK, then help found ARK), it fanfared a whole range of issues, from pollution and waste to food poisoning and loss of countryside. But its most eye-catching element was a list (as it turned out, a highly inaccurate one) of sixteen British wildlife species reported to be 'threatened with extinction'.

Public response to this part of the story far outweighed all the rest of it. A precedent was being set in media perceptions of what made newsworthy environmental copy and what didn't. There were, of course, deep-rooted precedents for animal welfare ideas, that had been closely tied in with liberal humanitarian concerns since at least the eighteenth century. But it could be argued that the 1970s language of threat and endangerment paraphrased subliminal fears for its own skin on the part of a public still on 'four-minute warning' alert against the threat of a nuclear holocaust.

Whatever the subliminal agenda might be, FoE soon realised that in endangered wildlife issues they had a tiger by the tail. The battle against fashion furs was also fought in the High Street to increasingly high-profile effect by supporters of FoE and by other groups with a more single-issue focus on animal welfare. For the first time, protesters carrying gruesome photo placards of slaughtered wildlife took to picketing fur retailers and barracking would-be customers.

The fur trade and allied businesses like leather goods retailing tried at first to bluster their way out of what they had not yet realised was a no-win situation. The International Fur Trade Association announced at the end of 1971 a self-imposed ban on using seven endangered species, as a move to forestall legislation. FoE was able to prove that the ban was ineffective. The trade magazine *Fur and Leather Review* pronounced that killing big cats and other wildlife was part of a natural selection process 'ordained by God' to boost the genetic heritage of hunted species. This plea was rightly ridiculed: weaker specimens with imperfect coats did not interest fur hunters in the least. Some fashion manufacturers, however, tuned in to the notion that, by marketing synthetic 'fun furs' as humane alternatives to animal pelts, they could score Brownie points with newly sensitised consumers and make a High Street killing. It was the first flush of a yet-to-come 'green consumerism' vogue.

Over the next twenty months, Pedley helped FoE draft a parliamentary Bill for an Endangered Species Act. It would be introduced through the House of Lords, by Lord Wynne-Jones, in December 1973. Although the Government opposed it, the Bill received strong support from the House. FoE eventually withdrew it when the Government was pressured into introducing its own Endangered Species Bill.

The moral of the tale in terms of its significance to pro-environment advocacy had been learned well in advance of its legislative conclusion. It was that media coverage got across messages that won the hearts and minds of people. Popular backing from ordinary voters might then prompt the politicians' will to move legal mountains and shift established power structures that allied big business with Government.

The UN Conference on the Human Environment intervened between this revelation and its acid test in Save the Whale campaigning; although it was at Stockholm in 1972 that the whales and whaling controversy first surfaced into truly international focus. One of the items on a side agenda at Stockholm was a proposal to install a ten-year moratorium on all commercial whaling, to enable researchers to establish reliable population figures for the main species used in trade. A total ban was seen as out of the question: the half-dozen nations most involved in whaling would have vetoed it out of hand. The moratorium was an exercise in the art of the possible and FoE, recognising it as such, backed it with zeal at Stockholm. In the end, the resolution went unopposed.

FoE had not been the sole initiator of this debate. Key advocates such as Roger Payne in the USA and Sidney Holt in the Fisheries Section of the UN Food and Agriculture Organization had already blown the whistle on declining stocks of great whale species, blowing loudest for the greatest of them all, the blue whale. Many new and established conservation groups then took up the cause.

Obsessive commercial exploitation throughout the nineteenth century had reduced whale populations in the North Atlantic. Remaining stocks were mostly concentrated in the Southern Ocean around Antarctica and along North–South migratory routes off the Pacific's edges. The whaling fleets that pursued them had grown in numbers and in their command of modern tracking and killing devices and floating factories for processing dead whales into commercial products. These could be stored under refrigeration without need of frequent returns to port. It was a recipe for extinction that could profit nobody in the long run; if whalers ran out of whales, where was their livelihood?

Whales formed, moreover, a key part of a hugely complex biological maze of food and energy chains. The survival of fish and other marine life rested on that system's stability. The FoE campaigners, originally fired by passions and by a vision of whales as a symbol of ethical understanding, employed such utilitarian and scientific arguments for saving whales more and more.

The whaling industry found some of these arguments easier to fend off than others. But its answer to the case that it was, so to speak, 'killing the goose that laid the golden eggs' rested on home-cooked figures of kill and stock levels that Holt and others had been able to show were way off the mark in several instances. The real state of play was, they felt sure, far more critical than industry estimates implied. The world's blue whale population was, for instance, thought to number fewer than 2,000 individuals, less than 1 per cent of natural levels. Unless further misguided exploitation was halted, extinction was written on the wall.

Figure 8 Legal adviser David Pedley (centre) and colleagues demonstrate in London, during the 1972 IWC meeting, when it was feared member states might not all observe the Stockholm moratorium

FRIENDS OF THE EARTH
SAVE THE WH
WHO NEEDS SPERM OIL?
Happy Christmas?
BAN SPERM WHALE IMPORTS
HAND IN GLOVE WITH THE WHALERS ?

Backed into a corner, even the most dyed-in-the-wool whaling states agreed at Stockholm to call a halt while definitive numbers could be established, to inform the setting of quotas so kills could be limited to a level that natural cycles of regeneration might replace, a sustainable harvesting regime. There was a lingering anxiety. The Stockholm Agreements were non-binding 'soft law' pacts. Would the club of whaling nations, represented by the International Whaling Commission, toe the line once the UNCHE publicity caravan had moved on?

It was with this nagging worry in mind that Christine Stevens contacted FoE in London in 1972 to alert them to the possibility that some IWC states might backtrack on the moratorium at the Commission's first meeting after Stockholm, to be held in London later that year. Graham Searle and his colleagues were quick to see an opportunity to make themselves conspicuous in the whale's cause.

Nicholas Holliman (Jonathan's brother) immediately began to draft a manual for a Whale Campaign, marshalling facts and arguments from Angela King and John Burton as well as from a regiment of allies outside the organisation. Graham Searle swotted up on the obscure wrinkles of the IWC's constitution, decoding procedures and chains of command. Angela King worked on cataloguing the products in which whale products were used in the UK. At that time the list included pet food, ice cream and margarine. Whale oil was also used in the motor, leather and watch-making industries as a fine lubricant or fabric finish.

The campaign advanced on several fronts and long outlasted the 1972 IWC meeting, which turned out to be, if not trivial, a relatively minor skirmish. During the period when the Commission met, FoE hired quarter-page spaces in the *Times* to run advertisements appealing to Britain to continue to back the moratorium and, beyond that, to introduce new legislation banning all trade in whale products in the UK.

An impressive list of public figures signed their names to the ads but HRH the Duke of Edinburgh, by now President of World Wildlife Fund, was at first a conspicuous absentee. Richard Sandbrook recalls:

> When we started Friends of the Earth, the 'enemy' was WWF, or rather the Establishment it represented, the 1001 Club and all that. They were pretty anti-us in those days, too. I remember the first time we ran the whale ad we got a call from Peter Scott to say: Why haven't you asked HRH to sign it? That was a day! We knew we'd arrived in that field at least.

Another triumph was scored on the showbiz front. David Bowie, a rising star then freshly metamorphosed into his space alien Ziggy Stardust *alter ego*, was persuaded by Angela King to dedicate a performance at the Festival Hall to the Save the Whale campaign. This was his first ever London appearance, netting huge publicity and a benefit, handsome in those days, of £2,500. Bowie was pictured in the programme as Ziggy, travelling from 'somewhere east of Mars' astride a giant whaling harpoon. His mission to Earth was described as a bid 'to Save the Whale!'

Figure 9 David Bowie/Ziggy Stardust benefit concert poster, 1972

Things were stirring out on the streets, too, with a feisty demonstration during September in Trafalgar Square by a chequered alliance of groups, coordinated by FoE, that had joined forces to form a 'Whale Fan Club' for the occasion. Several thousand supporters attended the rally and similar events were to follow in succeeding years, including an FoE spectacle that involved towing a giant inflatable model blue whale up the Thames to the Houses of Parliament. The whale, named Peter, ignominiously sank. Yet the media interpreted this foul-up as an intentional gesture.

Alongside all the fun and games, a strictly poker-faced lobby was being pressed in the corridors of power by FoE, using the Whale Campaign Manual (acknowledged on all sides as an accurate and credible contribution to the debate) to prove its credentials. Parliamentarians took the lobby and its public manifestations seriously. In due course these combined pressures were to score their first veritable hit.

In March 1973, in response to growing pressure and consumer boycott campaigns, the Government announced that it would impose a total ban on the import of all products from baleen whales, the filter-feeding as distinct from the toothed kind. That left the sperm Whale, traded for its oil and spermaceti (an expensive glandular substance used in perfumery), out in the cold. But, crucially, trade affecting blue and other highly endangered whale species had been blocked in what was once a key marketplace for whale products. It wasn't a total breakthrough, but it meant a lot.

Britain also stood firm in its support for the moratorium. Meanwhile, however, just as Stevens had feared, those IWC nations that had the biggest whaling fleets were doing their utmost to erode it, exploiting loopholes like a limited 'science kill' clause to allow researchers to collect life-cycle data and a provision for small-scale traditional whaling. No matter how outspokenly their actions were objected to inside and outside the IWC, whalers from Japan, Norway, Iceland and elsewhere continued to kill whales without fear or sanction, pleading these loopholes as their justification. The killing went on in mid-ocean. Who could do anything to stop it, short of waging war at sea?

As it turned out, there was an answer. But it wasn't within the repertoire of FoE, nor that of WWF or IUCN, despite their upmarket business connections and friends in government. It lay in the offbeat buccaneering tactics of what many then regarded as a lively yet marginal group of Canadian peaceniks and oddballs called Greenpeace.

Pete Wilkinson, who was soon to leave FoE to become a Greenpeace activist, can clearly recall the impact the Greenpeace phenomenon had on his own outlook.

> I can remember when Friends of the Earth did the first Save the Whale manual, all ninety-eight pages of it. I spent almost a week running it off on one of those old Roneo duplicators. We did the subject proud: it was a *tour de force* in terms of its analysis of whaling issues. But when later on I heard all these stories about Greenpeace out there in their boats in the Atlantic I thought Christ, yes, that's the way to go.
>
> I thought, the manual might raise a few eyebrows here and there. It might influence a couple of minor players. But it's not going to change the Russians or the Japanese or the Spanish or the Icelanders. What these guys were doing was much more in line with the way I like to approach issues. You know, once you've done your science and said this is the argument, what are you going to do about it? You've got to take it further. I began to feel Friends of the Earth was more of a talking-shop, more into analysis than action. When Greenpeace came along, I was immediately anxious to get involved in action-oriented work.

A time would come when even Greenpeace would fail to satisfy Wilkinson's definition of an action-oriented outfit. But few could fail to sense the allure the group radiated as it first took to the high seas with superbly insolent daring and style.

Though some of its earliest activists could lay claim to involvement in environmental actions from as far back as 1969 (the same year FoE was set up by David Brower in the USA), Greenpeace's first action under its definitive name and colours was an abortive protest voyage in the Bering Straits in 1971, to register concern over the US military's plans to conduct H-bomb tests on a barren rock in one of the most earthquake-prone regions on Earth.

In 1964 a major earthquake had hit this zone, wreaking destruction and death in Alaska and causing tidal waves as far away as Hawaii. Yet Amchitka Island, which lay smack on the geological fault where the earthquake originated, was chosen as the underground test site for America's nuclear arsenal. Appalled by this foolhardy plan, former US navy diver and marine scientist Jim Bohlen planned to take a boatload of young anti-war activists to the Aleutians, to draw attention to the risk the tests ran of causing further catastrophes. This idea of non-violent 'bearing witness' protest was borrowed from fellow Greenpeace tyro and devout Quaker Irving Stowe. Bearing witness to moral injustice had always been a key Quaker article of faith.

Stowe reminded Bohlen that a group of American Quakers had attempted a similar protest years before, in 1958, at Bikini and Eniwetok atolls in the Pacific, used by the USA as H-bomb test sites (and the origin, by the way, of the fashion term 'bikini', alluding to the shock effect of that anti-garment). The Quakers' peace voyage had no such effect. Their boat was confiscated and they were arrested in Hawaii before they could reach the test sites. Bohlen thought his action could avoid that fate by sticking to the neutral territory of the high seas and by mixing the nationalities of the crew so that any action taken against them would inevitably cause a diplomatic backlash. That way, even if the tests went ahead, the Pentagon's folly would still be exposed.

The group had an emblem, combining the universal symbol for ecology (the Greek letter *theta*) with the CND peace emblem, but they needed a name. A member of the group, Bill Darnell, came up with the verbal equivalent of this pair of images, Greenpeace, and it fitted. The protest vessel in which the group set sail in September 1971 was a rustbucket ex-trawler, the *Phyllis Cormack*. But the new name and double emblem were proudly flagged on its foremast. Greenpeace was launched.

It still had to face its sea trials, however, and from the start that maiden voyage was a rough ride, dogged by engine failures and seasickness, foiled by spoilsport tactical postponements of the weapons test and hauled off course time after time by fierce storms and a faulty compass. On its way out, however, the crew of the *Cormack* had one encouraging break when they put in for repairs at an island off the Alaskan coast and were fêted

Figure 10 Crew of the *Phyllis Cormack* return after the first Greenpeace trip to Amchitka to protest against nuclear testing, 1971 © Greenpeace/Keziere

by a group of Native Americans of the Kwakiutl nation. One of the crew members, journalist Bob Hunter, later recalled that this encounter seemed to clinch another vital element in the Greenpeace philosophy.

Many in the group, says Hunter in *The Greenpeace Chronicle*, had felt a pre-existing 'vague affinity' with Native American culture and its spiritual traditions, rooted in nature worship. It so happened he had brought a book about Native American mythology along on the voyage. It included an account of an ancient prophecy reminiscent of the biblical Ark and Covenant legend. Hunter passed it around with a sense that the finger of fate had marked its place. The red and white nations, the prophecy ran, would rally together round a band of wandering heroes who would save the world from a devastating era of plague and disaster. The name of these legendary rescuers would be the Warriors of the Rainbow.

Hunter's romantic account of this revelation has to be seen in the context of his own mission as an image maker who, as he later cheerfully admitted, had studied McLuhan and was curious to see how 'the medium is the message' might translate into 'media war' on behalf of an ethical project. But the brave sincerity of the Greenpeace founders and their commitment to a better world was never in doubt. What was in doubt during the Amchitka voyage was whether they would ever get there or back.

They finally got close but were arrested on a technicality by US coastguards, who informed them they should have reported their arrival in US waters off the island of Akutan, where they met the Kwakiutl, within twenty-four hours. They were told to return to Alaska to clear customs. As they headed back, the order was given to detonate the bomb test device.

They hurriedly transshipped to another vessel, sent after them by supporters in Canada, which they met on their way towards the mainland. It was a faster and more seaworthy craft than the *Cormack* but not fast enough. The test was over and done with before they got anywhere near Amchitka. They sailed home in a state of bitter frustration and depression. Yet on their return to Canada they found that colleagues on shore had been rushed off their feet by the media response to events. Their adventure and the issues behind it were hot news. The race had been lost but the laurels were emphatically theirs.

Interaction with existing pro-environment groups was crucial to the birth of Greenpeace. Like David Brower, Jim Bohlen had formerly been big in the Sierra Club and it was the Club that put up most of the funds to fit out the *Phyllis Cormack*. The group's new flagship, promptly rechristened the *Rainbow Warrior*, was purchased with a discreet cash gift from the World Wildlife Fund. Later on, the Greenpeace operation in North America would splinter over questions of power and principle. New Greenpeace chapters emerging in Germany, then in Britain, would take up the reins while the US factions slugged it out. The first Greenpeace Germany activists were former FoE activists; Greenpeace UK later harvested talent from the same seedbed. Later still, this compliment would be returned by crossovers in the other direction. The point was, few in the 1970s saw these two pressure groups as serious rivals.

A press announcement released in the UK by FoE in 1972 flagged a new Greenpeace act of witness against nuclear weapons testing, this time on Mururoa Atoll in the South Pacific. The press release, issued in the name of a peace coalition formed by CND and other bodies, offered friendly expressions of concern and solidarity.

David McTaggart, a Canadian business tycoon who had grown sick of the rat race and sailed off on an indefinite tour of the South Seas, learned in New Zealand that Greenpeace was looking for somebody to re-run the Amchitka action at Mururoa. The French were planning to explode aboveground test devices there in June 1972, the very month when UNCHE got underway at Stockholm. McTaggart got a crew together and they sailed his ketch, the *Vega*, into the test zone, coming close enough to the atoll to see the balloon that held the test device aloft.

They brought the *Vega* nearer but within minutes the frail craft was mostly matchwood. It had been rammed by a French minesweeper, *La Bayonnaise*, and was in imminent danger of sinking. The demonstrators had no choice but to come aboard the warship while it towed their shattered boat out of the area. Later, the French media published clandestinely shot photographs of the Greenpeace crew eating with French naval ratings on board *La Bayonnaise* as if at amicable ease with them. France denied aggression, blaming the damage to the *Vega* on 'clumsy handling by an inexperienced crew'. The lesson wasn't wasted on McTaggart: two could play at the media game.

Next time McTaggart braved the French navy off Mururoa, in August 1973, he came armed with a still camera and a video camera. When naval commandos came aboard and beat McTaggart senseless they also ditched the video camera. But the still camera, operated by crew member Anne-Marie Horne, captured all. This time around, when the French denied the use of violence (by their account McTaggart's injuries were caused by a fall), Greenpeace was able to publish evidence of their duplicity for all to see, in dozens of newspapers around the world.

These were just the opening rounds of a needle match between Greenpeace and France over nuclear issues that would go even to the death. In 1985 the *Rainbow Warrior* and crew member Fernando Pereira were blown up in Auckland harbour by French secret agents. For now, however, Greenpeace was jubilant. It had hit on a non-violent campaigning formula that promised to bring their activities and the issues they confronted right to the forefront of the world and media stage. All it took was four ingredients that could work under all sorts of circumstances: a David, a Goliath, a boat and (above all) a loaded camera.

With the appearance of Greenpeace on the scene, the direct action end of the spectrum of pro-environment pressure-group activities would shortly shift to become the centre of attention. Though few realised it at the time, this shift would have profound effects on the ways FoE and other land-lubbers acted and saw themselves from now on.

FoE had established itself as the senior force in 1970s pressure-group environmentalism in Britain, though there were few big rivals for that title. In terms of front-page media coverage, FoE was on a roll. It persuaded the Government to pass an Endangered Species Act. In the long run, FoE's innovative consumer campaigns, combined with astute political lobbying, achieved a total ban on all whale products, including sperm whale products, in the UK. With the passing of the 1982 Whale Directive, a similar ban was won Europe-wide. It had kept up a steady pressure on the IWC, too, fighting every new inroad into the moratorium with lucid counter-arguments backed by countrywide demonstrations and increasingly skilled media handling. And the campaigns and personalities so far mentioned are but a few among many which could be highlighted here.

None the less, things were changing. Greenpeace soon put its novel tactics to the test in the arena of wildlife issues, first as an opponent of seal-pup culling in Newfoundland then in high-risk confrontations in remote waters with moratorium-busting whaling fleets. And gradually – and deliberately – FoE yielded the hot spot of high-profile, heart-in-mouth campaigning for exotic forms of wildlife to the Direct Action derring-do of Greenpeace. While Greenpeace worked on international issues, FoE opted to scale down the whales campaign and put more effort into championing Britain's own wildlife.

FoE drafted a Conservation of Wild Creatures Bill, which became the 1975 Conservation of Wild Creatures and Wild Plants Act and was later incorporated into the 1981 Wildlife and Countryside Act. King worked on the campaign, initiated by Poole FoE, which in 1978 won protection from hunting and harassment for otters in England and Wales. And she carried out the detailed research into the decline in British wildlife habitats which provided the background for the 'Paradise Lost' campaign launched in 1980.

So whilst Greenpeace's crews of anonymous heroes took to the high seas, FoE continued to clock up an impressive series of victories, as well as moving the debate from focusing on charismatic animal species to an appreciation of the importance of habitat protection, using the traditional FoE methods of high-profile media and public awareness campaigns backed up by research and political lobbying.

What caused this subtle shift in tactics? Opinions vary wildly on this score, but it has always touched a tender spot with FoE campaigners. Czech Conroy recalls,

> Public opinion and media perception of what was newsworthy had changed over the seventies. The Schweppes bottle dump was big news in 1971, but if we'd done it in 1981 it would probably have been completely ignored. Come the late seventies/early eighties, to get coverage of a media stunt you had to do something more dramatic, and that often meant breaking the law. Greenpeace did this. Friends of the Earth's principle that it would always act within the law limited the range of options for us.

Established conservation groups had also taken a new lease on life from the new popularity these issues now commanded, thanks in large measure to the work of FoE.

Over the next few years, the development of FoE would be more and more tied in with questions about technology's role in society, establishing it as the group, more than any other, that worked on what would now be called a 'sustainable development' agenda. There would come a growing fixation on a science-based 'systems approach' to understanding the profound connections between everyday human lifestyles and the living planet. These sea-changes would also spur strenuous bouts of change and controversy within FoE itself.

If not the first, the most notorious manifestation of an altered outlook within FoE would be a series of intensely fraught confrontations over that awkward offspring of the Bomb, nuclear power. These disputes would fetch up in near-debacle at a place which is no longer on any map, though almost everybody remembers its name. Some things never go away.

8

Walls have mouths

The Zulu addressed himself to Sir Henry, to whom he had attached himself. 'Is it to that land that thou wouldst journey, Incubu?' he said, pointing towards the mountains with his broad assegai. I asked him sharply what he meant by addressing his master in that familiar way. Incubu is a native word meaning, I believe, elephant. It is very well for natives to have a name for one among themselves but it is not decent that they should call one by their heathenish appellations to one's face. The man laughed a quiet little laugh which angered me.

– H. Rider Haggard, *King Solomon's Mines*, 1885

Changes within the environmental movement and in the character or management of Friends of the Earth were not – of course – the only kinds of change that determined the course of environmental action and debate in the Britain of the 1970s. As always, those events had mainly been driven on by changes in the complexion of political affairs, by stories that made news regardless of anything pocket-sized pressure groups did, and by undercurrents in the culture of the day.

A shortlist of headline events that would directly or indirectly alter the political, social and economic odds affecting the environment in Britain and the world during the decade and long after would have to include the discovery of deposits of oil and gas in the North Sea from about 1970. At around the same time, Poland's shipyards were paralysed after a massive workers' revolt in Gdansk, while Vietnam protests exploded in 1971 as hostilities spread to Laos and Cambodia.

In the same year, West Germany's Willy Brandt was awarded the Nobel Peace Prize for his espousal of Ostpolitik, a brave attempt to bridge the East–West divide in Europe. Britain joined the European Economic Community (now the European Union) in 1973. But neighbourly relations in other directions hit new lows with 'Cod Wars' over fishing quotas off Iceland and shock IRA bombing campaigns in London and elsewhere on the British mainland, peaking in 1973.

In 1974, the USA finally withdrew its army from Vietnam. But an oil price hike imposed by the OPEC cartel caused a global economic crisis.

It led in Britain to the Three-Day Week which scuppered Edward Heath's Tory Government, reinstating Labour and Harold Wilson. It also confirmed some of the ideas in the enormously influential *Limits to Growth*, a doomwatch compilation by a research team at Massachusetts Institute of Technology on behalf of another global panel of pundits, the Club of Rome, recently published in the UK by the Earth Island Press, a publishing house set up by David Brower. *Limits to Growth* was the first convincing presentation of hard figures and measured forecasts of economic and population growth, compared to what was known of natural resource limits. The oil crisis was evidence of its arguments that continuous and increased consumption (hardly questioned before) could affect the quality of daily life for ordinary people.

In the same year, an explosion at a chemical plant at Flixborough on Humberside killed twenty-nine people. Richard Nixon was impeached and disgraced following the Watergate scandal. India tested its first nuclear device against the grim background of a smallpox epidemic in which at least 20,000 people died in four months.

The summer of 1975 saw the height of an imported Dutch elm disease pandemic in the British countryside which was to alter the appearance of large tracts of rural England. UNEP set up shop in Nairobi and in 1976 sealed the first 'bioregional' environmental pact, the Barcelona Convention for the protection of the Mediterranean environment, model for much global environmental lawgiving that was yet to come.

The optimism of Barcelona was dimmed, however, when the industrial suburb of Seveso in northern Italy was contaminated by a leak of the pesticide by-product dioxin. Many local people were disfigured by a hideous skin condition, chloracne, with no known cure.

In April 1976, the oil rig Bravo blew up in the North Sea and the Royal College of Surgeons published data showing that each time a cigarette was smoked it cut the user's life expectancy by 5.5 minutes. In 1978 the oil tanker *Amoco Cadiz* sank off the coast of Brittany, causing massive oil slicks that wrecked seaside resort beaches and nature reserves. The Green Party was officially formed in Germany. A million British workers were unemployed and many of the rest were on strike. The Winter of Discontent put Labour out in the cold in 1979 and in 1980, the first full year of Margaret Thatcher's twelve-year incumbency, yet another oil platform collapsed in the North Sea, killing 100 crew. And a US base for Cruise tactical nuclear missiles was established at Greenham Common.

There was more, much more, but a glance at this fragmentary list is surely enough, in hindsight, to make the point that as the 1980s approached the pressure was on groups like FoE to make sense of these swings of fortune. They went with exaggerated mood-swings in the culture at large, from self-absorbed escapism or apathy to frustrated spleen directed at orthodox politics – or orthodox protest. Some might say the 'we generation' was giving way to the 'me generation'.

Maybe so, but were things that simple? Maybe both outlooks were there all along but the balance between them shifted. And the different ways popular culture processed environmental issues held up a mirror to this split identity.

The late-1960s sci-fi movie *Soylent Green* (based on a story by Harry Harrison) pictured a future landscape entirely occupied by derelict urban sprawl, an overpopulated 'dystopia' where violence is the norm and where gridlocked automobiles lie abandoned everywhere. A giant food monopoly rules the roost, surreptitiously converting human remains into carefully disguised green food pellets, having run short of genetically engineered soya and lentil feedstocks. The lonesome heroes of the tale quixotically investigate and daringly expose this atrocity but by now their fellow citizens don't give a damn.

On a different tack, Ernest Callenbach's best-selling novel *Ecotopia*, which came out in 1974, portrayed the antithesis of *Soylent Green*, a green republic formed by young dissidents in a fortified enclave on America's West Coast, where government is modelled on natural ecosystems; a caring, de-marketised, vegetarian democracy defending a self-sufficient resource base. But in the end, the forces of Mammon smoke them out.

Both stories hinged on hopes of Utopia that neither militant action nor idealistic consensus could deliver. In the late 1970s, UK pop culture shadow-boxed between extremes, too, the in-your-face nihilism of Punk Rock alternating with the clubbable retro-pedalling of the New Romantics.

Compared to the anxieties evident in society and events at large, the flurries of soul-searching and upheaval that FoE underwent in the late 1970s and early 1980s were storms in teacups. But they sprang at least in part from diverging attempts to make sense of what was going on out there. Some insiders wanted the organisation to show more streetfighting nerve, some wanted more soul and creativity. Some wanted to fixate on global issues, some on national politics, others on local-level lifestyle issues.

A few mundane changes, more or less at a housekeeping level, had intervened since the organisation was set up. The first of these had been a fond farewell to King Street, with its piles of vegetables, bawling porters, all-night pubs and constant toing-and-froing of produce trucks. The group knew it had overstayed the kindly welcome extended by Ballantine Books but it had outgrown the space in any case. New, larger premises were found in Soho's Poland Street, offered on fixed-term loan by a philanthropic Quaker foundation, the Joseph Rowntree Social Services Trust. Friends of the Earth Ltd moved there in 1972 and it remained their headquarters for ten years.

Mainly through the efforts of Richard Sandbrook and Barclay Inglis, the organisation had acquired a functional Board of Directors to advise on policy matters and to steer its finances and management. Far from settling into a passive role, the Board was in the thick of it from the beginning. Sue Clifford found the Board experience a rollercoaster ride.

> As Directors we were always being asked to mediate in what was going on with the staff or deal with constant financial crises. There were always big questions about the directions people wanted to go. One was aware that one was in something that was growing as an idea as well as an organisation. So there was never a quiet time, ever. There wasn't one meeting where one just sat and relaxed. There was always some wringing-out of one's brain or one's compassion!

Another Director (and later Chair) of the Board was Charles Levinson, a showbiz lawyer who became interested in FoE after helping arrange the David Bowie concert. He knew most of the important figures in pop music and, through him, many of them began to get involved in pushing the FoE agenda. The celebrity supporter roll-call included, at one time or another, Andrew Lloyd Webber, Mike Oldfield, even the Sex Pistols.

Searle had been running things for a little over two and a half years when the move from King Street was completed. Sandbrook joined the Poland Street payroll in September 1973, having worked on a voluntary basis up till then. When he announced to colleagues at Arthur Andersen that he was going to work for FoE, everyone thought he was mad. 'One partner said: 'What? You're going to work for the Flat Earth Society?' But by this stage we'd grown up a lot.' He signed on officially as Company Secretary and 'money person'.

One of the most pressing tasks awaiting him was to sort out the local group relationship. Even in 1972, not content with an occasional Campaign Manual coming down from the mountain, the groups had begun to lobby for faster feedback from the centre and more say in policy. Richard Sandbrook remembers 'the local groups began to say: hey! Where's the democracy in all this? In their eyes we were a club, a self-appointed hierarchy of mates, and the Board were mates too.'

Sandbrook invited Tom Burke of Merseyside FoE to come to London to coordinate the local groups network and to codify its relationship with the centre. The invitation was accepted. Burke had been outspoken at an inaugural local groups rally held earlier that year and since then had circulated a paper calling for a more structured approach to liaison between groups and centre. His own local group was one of the best organised in the country, though by no means the largest. 'We were very rumbunctious up in Liverpool,' says Burke. 'I was 24 years old, I think I was the oldest person in the group by four or five years.'

Burke was teaching Liberal Studies to university students when he grew aware of FoE through the Schweppes ad campaign. He came to London and met Graham Searle, who encouraged him to form a local group and gave him leave to use the FoE name. He recalls,

> There were five or six of us originally, then in due course about twenty, a very vigorous group. We did photo exhibitions, actions on packaging. Two of us decided to leave our jobs and go for full-time campaigning. We had to figure whether we could make enough to pay ourselves a salary or not, to do what we wanted to do. We had an office of sorts. It doesn't sound

> much now but then it was fantastically ambitious. It was the central office's failure to make use of what we saw as our tremendous potential that got to us in a while.

Today, Burke insists that he and his fellow local group members had 'no great idea that we wanted to stir up some sort of revolution. We didn't want to press our view of how things should be run on the principals in London. These were mythical figures, magical people we were in awe of,' he reminisces. Nor, says Burke, did they cherish illusions of themselves as a green counter-culture.

> This was the early seventies, remember. Ideas like Green weren't in anyone's mind. There wasn't any theory, nothing to theorise about, just bad things happening. Friends of the Earth was militant compared to groups like CPRE, in the conservation tradition, though the Conservation Society had just done quite an aggressive protest about landfill sites.
>
> But Environment was driven out of a population, resources, pollution tradition. It was about the future as distinct from the past. I think if anything the real mobilising force was anti-authoritarian feeling, and lack of belief in the political process. The idea of turning to politicians to do something was just silly. Friends of the Earth gave you a way of engaging directly against things. It was an age of protest, protest was a legitimate thing. Theory came later.

Burke settled in at Poland Street and began sorting out a role for himself.

> When I came in nobody seemed to know what I was supposed to do. I just got on with it, got involved in everything. Then I discovered I could write so I ended up writing lots of things, including the first Newsletter for local groups. I just saw it as the political job of mobilising the groups to support the campaigns. I also saw it as giving me a lot of say in what happened. I was the voice of the local groups coming out of Poland Street.

A few months after Burke arrived Graham Searle announced that he was leaving the organisation. Richard Sandbrook cuts a long story short: 'Graham went off to New Zealand saying: Well, you're here now and Colin Blythe, you sort it out. Colin Blythe and I had a tussle and eventually he said, all right, you run it. Which is how I became the second Director.'

Searle travelled Down Under and got back in touch with his love of wilderness by writing *The Rush To Destruction*, a ringing polemic about the destruction of New Zealand's Pacific Drift rainforest. He would return from this sabbatical to figure prominently on the research side of FoE in the future. But for now the Searle era was done.

Walt Patterson was still around, however. Titanic struggles were impending over nuclear power issues, and in these he would soon be helping establish FoE as a core contender. Straight after leaving Stockholm and signing up with FoE, Patterson had gone to America at the invitation of David Brower. Brower wanted him to be in Washington with Richard Wilson to do another two weeks of *Eco* issues, billed as the title's

Volume Two, reporting nuclear safety hearings which were being held before a Congressional Committee in August 1972.

For Patterson it was a copy-writer's baptism of fire. He had to research and write a 2,000-word article every forty-eight hours. 'The conference at Stockholm was fun but I didn't have to do anything really heavy,' Patterson relates. 'But Washington was quite a different proposition. We hit every key nuclear issue on the head in that series. So when I came back to London in the autumn I was saturated with nuclear issues. But there wasn't anything happening in the UK on that score.'

On 15 October 1973, however, Peter Rodgers ran a front-page piece in the *Guardian*, saying that the Central Electricity Generating Board (CEGB) was planning to abandon British-made gas-cooled nuclear reactors and import American Pressurised Water Reactors (PWRs) in their stead. Initially developed to power nuclear submarines, PWRs were the most compact reactors available at the time, requiring less elaborate and expensive infrastructure to install than other types then in use. But their safety prognosis was dubious.

Rodgers pointed out that their design depended on maintaining a steady flow of water under pressure round the reactor core to prevent it from overheating. But whereas systems that used gas or other coolants were reasonably forgiving if the coolant supply stopped circulating, allowing time to shut the core down at a fairly lazy pace, PWRs required an instant shutdown response to avoid meltdown if the water supply failed through leaks or obstructions. They were, some experts claimed, too 'frisky' to trust as a scaled-up power generation system.

Such warnings would re-echo in 1979 when a PWR at Three Mile Island, near the US city of Harrisburg in Pennsylvania, was the cause of the biggest civil nuclear accident so far. In the mid-1970s, however, nuclear technology wore a well-buttoned cloak of respectability and bland secrecy. A BBC radio contact of Patterson's called him to ask if he'd seen Rodgers' article. Were PWRs such a bad idea? Yes to both, Patterson said. He was asked over to Broadcasting House to talk about it on radio.

The interview went out on the air the same night. Patterson says,

> The following day the Director General of the BBC got a furious telephone call from the Chairman of the National Nuclear Corporation, saying: who the hell is this Patterson? What do you mean by letting him badmouth our technology on national radio? The producer was able to say, well, we asked Sir Arnold Weinstock – then Director of GEC and a Director of NNC – to come on the programme but he wouldn't do it.

FoE had just published the first edition of Amory Lovins' *World Energy Strategies*. Lovins followed this up by writing a half-page article for the *Sunday Times* on the safety issue. He and Patterson called a joint press briefing on the CEGB story. Patterson recalls,

> The place was packed. We thought we'd just be giving a few reporters a quick run-down on the main issues. We had every science and political

> correspondent from all the newspapers there and it went on for nearly two hours, it was like a seminar. They wrote up a whole lot of stuff about it.

Shortly after, the Commons Select Committee on Science and Technology decided they would hold hearings on the CEGB proposal, due to start in December 1973. FoE was invited to give evidence. Lovins and Patterson conferred. Says Patterson:

> Amory was all for going in, firing in all directions. I said we didn't have a constituency to confront nuclear power on all fronts. There's no way people in this country will get interested in that. If we want to win this campaign we have to focus on PWRs and forget the rest.
>
> What Amory didn't know and I did was that there was deep hostility on the part of the British nuclear establishment toward the Americans. In a matter of weeks they were up in arms against the American reactors, we could sit back and let them fight one another. That was one powerful lobby.

In December 1973, the Committee hearings took evidence from Sir Arthur Hawkins, then Chairman of the CEGB. Lovins and Patterson followed events in the Committee Room, waiting for somebody to shoot themselves in the foot. Sir Arthur obliged. Patterson recalls,

> He was just unbelievably arrogant. He had told the Committee sixteen months previously that he could see no need for further nuclear power plants, except possibly one plant for the following decade. Now he came in and revealed to them, almost in a subordinate clause, that what the CEGB was planning to do was order one or two stations of the American type every year for more than a decade – at least thirty-two PWRs.

They were going to be of what Hawkins called 'an off-the-shelf, bread-and-butter design'. Pressed further, he specified a 1300kW Westinghouse PWR. 'There was no Westinghouse 1300kW PWR,' says Patterson gleefully. 'There never had been. In those days, the largest one that had ever been started up was 1100kW.' Lovins and Patterson spent that Christmas writing a memo to the Select Committee, working twelve hours a day in close harness, going through all the literature that might illuminate the comparative safety of different reactor types. Patterson explains:

> The trouble was, most members of the committee didn't like American reactors but they didn't know why they didn't like American reactors. Their report came out in February 1974 and our memo was the first document given in evidence, saying that PWRs were an extremely bad idea and giving all sorts of reasons why. Which looking at it now turned out pretty well-founded.

The end result of the anti-PWR campaign was to help persuade the Labour Government to cancel the proposal to buy thirty-two reactors. Severe public expenditure constraints were undoubtedly the key factor in the decision. But it was now clear that the CEGB's grandiose expansion plans were a nonsense for energy policy and technical reasons.

The winter of 1973–4 was the time of the OPEC oil shortages which, combined with problems in the national gas supply, triggered the Three-Day Week. Energy was box-office. 'It got to the point where there was this bizarre business of big corporations advertising their nuclear power reactors in the daily press,' says Patterson. 'Two-page ads from General Electric and Westinghouse saying please buy our reactors.'

In June 1974 Patterson appeared as a witness at another nuclear inquiry – into proposals to build an unspecified number of reactors of unspecified type within an unknown timescale, at Torness. FoE's published evidence included technical proposals, which subsequently proved to be well-founded. However, the vague nature of the project made the inquiry a hard one to fight. Four years later, the Government announced that it was going to build a reactor, claiming that the inquiry had examined all the issues. This was untrue.

Patterson was in double harness, for he was working on North Sea oil issues at the same time. Though not involved exclusively in energy issues when all this began, he was soon *de facto* FoE's first Energy Campaigner. In the summer of 1973 a message came from FoE campaigners in Scotland, inviting Patterson to a meeting in Dundee to look at issues arising out of North Sea oil development. Other groups were involved, including the Conservation Society. At the Dundee meeting, they pointed out something the national press had not yet picked up on.

It was a plan to take over land for production sites for concrete oil platforms and in particular to take over a piece of land on the mainland by the Isle of Skye, at a village called Drumbuie. 'We decided this wasn't basically fair, that the Department of Industry and the construction industry and the oil companies should gang up on this little hamlet of just a few houses and a few dozen people,' Patterson recalls.

He travelled to the Kyle of Lochalsh with a suitcase full of documents about policy issues involved in North Sea oil development, particularly the plans for offshore platforms. He spent about two hours in the front room of the local chemist along with the schoolteacher, the postmistress and various other locals who were deeply worried about what was happening. He persuaded them that they had a serious technical case against the plan that was being adopted over their heads. But if they wanted to fight that case at an inquiry they had to be able to hire an advocate to represent them. They listened.

> In the end they said: 'Right, we'll do it.' And they put their own money forward, at least 5,000 pounds, quite an amount. So for the rest of that winter as well as running the PWR campaign I was also looking after the London end of what we called the North Sea Oil Coalition. In July 1974 we won the PWR campaign and also drastically scaled back the nuclear expansion programme from forty-one to just six gigawatts. In August we won the Drumbuie campaign, 'we' meaning the Coalition, it wasn't just us.

Not only had FoE arrived as a rated player in the nuclear arena, but in the Drumbuie process it had learned that public inquiries were winnable and that there was scope to play a worthwhile co-enabling role on the side of a local community defending a prized environment against energy Goliaths and the powers-that-be. The question of which was the more important lesson would not lag far behind.

In January 1975, two former site workers at the Windscale nuclear reprocessing plant died of the same, extremely rare type of cancer within twenty-four hours of one another. Alarm bells started ringing in the media, for Windscale was already infamous as the site of the world's first civil nuclear accident, a serious fire that broke out back in 1957. British Nuclear Fuels Limited (BNFL) was sufficiently set on its guard to hold a press conference at Windscale, the first since 1962.

Health issues dominated the questioning but Walt Patterson raised an unexpected query. Was it true, he asked BNFL's Chairman, that Britain was exporting weapons-grade plutonium processed at Windscale to Japan and Italy, neither of which had signed the Nuclear Non-Proliferation Treaty? Patterson well knew the answer was yes. Thrown off balance, the Chairman replied that it wasn't a question for him, Patterson would have to ask the Government. Having thus let the cat out of the bag about continuing and increasing ties between civil and military nuclear technology, FoE was coming out of the green corner as a heavyweight anti-nuclear contender when the next most crucial round began.

A report appeared in the energy trade press towards the end of 1974 announcing that BNFL was planning to build two huge new Thermal Oxide Reprocessing Plants (usually known as THORP 1 & 2) at Windscale. In May 1975, FoE produced a four-page mock tabloid newspaper called *Nuclear Times* written by Patterson. One of the two front page stories asked 'IS FISSION WORTH IT?' and tore the half-baked economics of CEGB and BNFL to pieces. The other was headlined 'WINDSCALE TO BE WORLD CAPITAL FOR RADIOACTIVE WASTE'. 'That was the article in which I made the "nuclear dustbin" analogy which then got picked up by the nationals,' says Patterson, 'from then on it just spiralled'.

Later that year, the *Daily Mirror* picked up the story and ran a front-page denunciation of the Windscale proposals under the banner headline 'PLAN TO MAKE BRITAIN WORLD'S NUCLEAR DUSTBIN'. This was the first of many counterblasts to the proposals from the media, which were growing increasingly alert to environmental stories and increasingly adept at handling them.

Within FoE meanwhile, a powerful campaign team had been assembled for the forthcoming nuclear showdown over Windscale. Czech Conroy, already a veteran local groups campaigner at the age of 26, became the

Figure 11 Windscale demonstration in Trafalgar Square, 1978

NO
TO WINDSCALE
NOW
FRIENDS OF THE EARTH
WE WANT 'SOFT' ENERG
WINDSCALE
cap de la HAGUE
SAME
FIG
NO JUSTICE FROM
PARKER
FRIENDS OF
WINDSCALE
BOMBS
THE EARTH
TAKE

team leader in January 1976 while Walt Patterson focused on research and Amory Lovins gave assistance during those months that he was in the UK. Many others were to join them as the campaign gained momentum, including a young energy specialist called Mike Flood who was to prove an expert anti-nuclear fighter over the forthcoming decade.

But at the start of 1976 an inquiry at Windscale was far from being a tangible reality. The energy team spent the entire year trying to force a reluctant Government to concede that an inquiry was necessary. It was only in December, after twelve months of petitions, demonstrations, media stories and Early Day Motions, that the Government backed down and announced that an inquiry would be held the following year. As Czech Conroy remembers, the simple announcement of an inquiry itself felt like a victory: 'It was a major achievement for us and it's probably not appreciated now just how ground-breaking it was. The Government really, really didn't want to have that inquiry.'

The date was set for May 1977 and the campaigners were thrown into a flurry of preparations. There was a barrister to hire (and brief), witnesses to organise and transport (some from abroad) and piles of documents to prepare, while the campaign pressure was still relentlessly applied in the outside world. Not only that, the team was still campaigning on its 'positive alternatives' agenda as well and local groups around the country were insulating pensioners' lofts and campaigning for better insulation standards for homes. They even managed to publish a report on renewable energy. But when May came round it was time to put aside distractions and focus on what was to prove the longest public inquiry in British history at that time. The 100 days of Windscale had begun.

Walt Patterson would like to stake the (not uncontested) claim that he pioneered the subversive use of the three-piece suit at the Windscale inquiry. 'When I joined Friends of the Earth everyone wore jeans and sandals. I said, if you're going to be in the lobby at Westminster or with the lawyers, you want to be able to get in there, say your piece and leave before they realise they've been talking to a weirdo! So within a year everyone was doing it.' At one point, he admits even to wearing a pocket watch with a fob chain. Whether or not the power dressing made a difference, FoE emerged at the very start of the inquiry as a force to be reckoned with in the wary eyes of the industry side, and with good reason. 'They realised we knew more about the issues than they did', Patterson claims. 'We were putting our case and cross-examining witnesses so effectively that at one stage we had a witness for BNFL so angry he was practically in tears. We had a very good QC, Raymond Kidwell, who made a great case.'

As the hearing progressed, members of the team began to feel an increasing confidence. The senior High Court Judge who headed the inquiry, Justice Parker, seemed to be willing to give their measured views a more than generous crack of the whip. He was even on first-name terms with the team if he ran into them outside the hearing sessions in the little Cumbrian town of Whitehaven. The judge seemed to have understood their arguments and Patterson was optimistic about the chances of a win.

Even the cautious Czech Conroy sensed that things were moving their way. 'I thought the odds were stacked against us at the beginning,' he remembers, 'but we put together a very creditable case. By the time that the inquiry ended, I reckoned our chances of victory at about fifty-fifty.'

After the inquiry was over, FoE concentrated on pushing for publication of the Parker Report, delayed until March 1978, and a debate in Parliament before any decision was made on whether or not to go ahead with the new plants at Windscale. But when it finally appeared the Parker Report turned out to be (Patterson later wrote) 'a numbing dismissal of every opposition argument', which 'could have been written without even holding the inquiry'.

Conroy also recalls the bitter taste left by Judge Parker's conclusions. 'What really stunned me was the incredibly one-sided nature of the report. We had really believed that our ideas would be accurately represented and objectively considered – even if the final judgement went against us. What we got was a completely one-sided report which totally failed to take any of our ideas on board.'

So what went wrong? Reviewing the first ten years of FoE in a *New Scientist* article in April 1981, Jeremy Bugler wrote:

> I still have fresh in my mind the winded astonishment of Friends members when they heard the news of the Parker Report. They were like men who had been asked to speak at a meeting for a worthy cause and found that while they were doing so, someone had picked their pockets. In my judgement, Friends of the Earth failed to realise that it was going before a hostile bench.

'Friends of the Earth was politically naive at Windscale', Bugler's article accused. 'But what is arresting, in trying to hold a ruler to the Friends, is that this naivety is accompanied by a commitment to politicking, accompanied by a march-them-up-the-hill belief in the political system and the trustworthiness of Whitehall, that has misled Friends of the Earth and the whole environmental movement.' A severe criticism indeed – and not wholly justified. Perhaps Bugler's article betrayed some of his own disappointment – for he was not the only media commentator who recognised that FoE had won the arguments.

Bugler's piece went on to draw unflattering comparisons with Greenpeace, which had just bounced the issue of seal culling in the Orkneys with an uncompromisingly emotive appeal to popular sentiment, combined with non-violent guerilla tactics. Tactics like these, said Bugler, were the way to go; to scurry and squeak around the corridors of power was the way to arrive in a mousetrap.

Unsurprisingly, members of the FoE energy team strongly disagree with Bugler's cheerless analysis. Says Conroy,

> It's true that we lost, but we gained a great deal in the process. There was a massive raising of public awareness about nuclear issues because of the daily media coverage of the inquiry and there was a big jump in the credibility of Friends of the Earth. They couldn't dismiss us as ecofreaks any more.
>
> It also established a significant precedent. The Government could never again consider a major nuclear development without holding a public inquiry first, and that in turn delayed the progress of the industry at Sizewell. BNFL knew that from then on it couldn't take public ignorance for granted and that it would have to argue its corner for every development.

Patterson also disagrees with the media critics: 'If we hadn't campaigned as we did the construction of the plant would have started in 1975. As it was they didn't even break ground there till 1983, by which time all the points we raised were fully borne out.' Nor does Patterson think that the verdict of the inquiry was necessarily a foregone conclusion:

> The turning point for me was that everybody else, once the inquiry really got under way, climbed on the bandwagon, everybody with an environmental philosophy wanted a piece of the action.
>
> If it had been left to us, there was no way that the anti- case could be ignored. I think that it was a successful tactical move by BNFL to invite everybody into the room so that they could get under each other's feet and clutter up the arguments. What happened, there was a huge smear of arguments ranging from the very well worked out to the wildly angry and exaggerated, which left room for Parker to do what he did, to tar everybody with the same brush and say it was all totally over the top, indefensible and so on, invoking everybody at the same time.

In retrospect, FoE's presence at the inquiry had done enormous good, raising the organisation's profile to a level it had never previously achieved and marking it out as the credible voice on nuclear issues. It had severely undermined the case for nuclear power, rattled a powerful but complacent industry and challenged the conventional wisdom that nuclear waste should be reprocessed. And, some twenty years later, the combined weight of two decades of campaigning paid off when the Government conceded in 1994 that there was no justification for building any more nuclear power stations in the UK.

Conroy is bullish in his assessment of the anti-nuclear campaign.

> I think it was one of the most effective campaigns that Friends of the Earth ever ran, especially when you consider what we were up against. Britain was a world leader in nuclear technology, with both civil and military programmes. To stop the expansion of the civil industry in only twenty years of campaigning was an incredible feat.

Final victory seemed a distant prospect in 1978, however, and FoE found itself both demoralised by the loss at Windscale and caught within a furious debate about whether, given the failure of legal methods, direct action must now be the way forward. A way that Greenpeace was happy to demonstrate with increasing success.

Local groups were unhappy too. Their numbers and professionalism had been rising inexorably and they had now reached a stage in their development when the old systems of semi-formal central control seemed antiquated and unjust. A new clamour emerged from outside London which was to turn up the temperature of a bitter power struggle already brewing at Poland Street. A sombre season was about to set in, when friendliness would sometimes be in short supply at Friends of the Earth.

9

The appliance of science

> The environment does not exist as a sphere separate from human actions, ambitions and needs. Attempts to defend it in isolation from human concerns have given the very word 'environment' a connotation of naivety in some political circles. But the environment is where we all live – and development is what we all do – in attempting to improve our lot within that abode. The two are inseparable.
>
> – Brundtland Report, *Our Common Future*, 1987

Friends of the Earth had entered the build-up to the Windscale campaign under Tom Burke's direction and with hopes building to a high. It departed from the Windscale fallout under a cloud and with a much altered management regime. Crucial changes at the top and in the London infrastructure had taken place since 1976. Initially, they had become associated with an impressive clutch of successful achievements. But by 1978 the new regime began to look increasingly rocky. By 1981 the knives would be out.

The main changes in the group's leadership began with the departure of Richard Sandbrook in 1975, though he was to remain an influential adviser for several years to come, and remains a Trustee to this day. He quit his primary involvement mainly so as to take up an offer to help establish a new think-tank, the International Institute for Environment and Development.

Other factors coloured his decision to go. As he candidly admits, 'I couldn't go on earning absolutely nothing or very little. It was very dodgy financially.' Though big gifts from celebrity well-wishers and donations from the public had grown steadily since the group's outset, there was never enough cash to go round.

Unlike more pyramid-shaped independents like WWF or Greenpeace, a subscription membership had never been the main source of the group's revenues. Local groups were self-governing and many contributed sums for particular campaigns. The system of pay parity at head office, whereby each staffer earned the same regardless of seniority, was admirably fair and helped the cash go further.

Some staff questioned the wisdom of retaining this system, however. Now that FoE was not the only fish in the sea, the parity rule could make it harder to bring fresh expertise in from outside. (The group had nevertheless managed to hook Joanna Gordon-Clarke, a former environmental policy manager for the oil and gas multinational BP. She replaced Angela King, who had left in late 1975, intent on country life and a more direct and less fraught interaction with nature.)

Sandbrook had also established rules and procedures for the group's financial conduct. None the less, nobody in Friends of the Earth Ltd or on its Board had any fund-raising qualifications or experience. Direct mail fund-raising appeals, a standard income-generating procedure for all the big environmental groups by the mid-1980s, were practically unheard-of at that stage. Finances never seemed to match the group's ambitions.

Shortly before leaving Poland Street, Graham Searle had done his best to get the organisation into healthier financial shape by dining with the Devil. At the playful invitation of society gambler and zoo-owner John Aspinall, he attended a private dinner in Belgravia along with some of the wealthiest industrialists in Britain, including Teddy Goldsmith's multi-millionaire brother, Sir James Goldsmith, and Lord Rothschild, owner of the banking group. Searle gave an after-dinner presentation about the aims and work of FoE.

What followed was later described by Searle as a 'bear-baiting' session where each of his fellow diners threw at him the trickiest question they could think of, which he undertook to answer off the cuff. If they felt convinced by what they heard, his listeners pledged to put a donation in a hat to be passed around at the evening's conclusion. The cheques in the hat totalled £14,000, more than the organisation's whole first year's expenditure. But such windfalls were few and far between and they weren't much of a way to run an ethical concern in any case.

When Graham Searle returned to London from New Zealand in 1974, no obvious role remained for him in the organisation. He therefore undertook to develop Earth Resources Research, the policy research arm of FoE that he and Sandbrook had founded two years previously. The two fell out over the terms under which Searle would be engaged and an undercover plot to unseat Sandbrook was hatched by a small caucus of staff members in the heat of the row. A face-saving deal was struck, however, and it was agreed that Searle would run ERR on separate premises.

He was to do just that for several years and to great effect, as ERR produced seminal policy papers on agriculture, energy, population, food, transport and other key issues. It meant, however, that Richard Sandbrook had been denied a cherished policy development role within the organisation, and it seems likely that this coup may well have hastened his decision to quit.

By his own account, however, it was a decision from the heart. 'The real reason why I left was that I'd met Barbara Ward and fell in love with her message,' he says. 'To work with her was incredible. During my last few months in Friends of the Earth I'd become more interested in the Rich World, Poor World dimension of things and in the international agenda.'

He was, he says, more than enthusiastic about prospects for the UN Environment Programme and what he calls 'UN-ery' in general, though he admits to feeling more jaded now. 'Realistically, things like oil pollution at sea were only going to be solved by the International Maritime Organization. And I was very keen on what was being achieved over whaling and fisheries by Sidney Holt at the UN Food and Agriculture Organization, another of those saints of the business.'

Sandbrook had tried to get FoE and similar independent groups into a stronger position to respond to global issues. In 1974, at an FoE International meeting in Nairobi, Sandbrook and others had tried to set up a Nairobi-based FoE International, on the doorstep of the UN Environment Programme. This had not worked, but did lead to the setting up of the Environmental Liaison Board, later the Environmental Liaison Centre, sited in Nairobi. Peter Hayes, ELC's first Director, was formerly Director of FoE Australia.

'We didn't want ELC just to be identified with Friends of the Earth International,' says Sandbrook, 'it was a wider party than that.' There was, however, an element involved of keeping groups like WWF International, which had what was seen as a rather too cosy relationship with the intergovernmental establishment, out of the act at the start, so that the Centre could operate freely as a ginger group and watchdog over the UNEP.

Back in the Poland Street office in 1975, there were two contenders to replace Sandbrook in the top job. Tom Burke pipped Joanna Gordon-Clarke to the post, and was to remain in charge until 1979, when Sandbrook invoked the three-year rule and engineered his departure.

Tom Burke's robust management style was in many ways a breath of fresh air for the organisation at a time when it seemed in in danger of dithering over the direction its campaigns should go. Burke was pragmatic:

> We campaigned on well-researched issues with achievable objectives, so all the campaigns were significant. We might select particular issues to concentrate on that stood in for much broader issues. But we weren't interested in arguing about those issues in public.
>
> We were interested in outcomes that would advance the broad cause. If you didn't deliver results, nobody would support you. So you chose the type of arguments that would deliver results. That was a very clear formula understood very well by all the twenty or so members of staff at that time. We all worked hard at it and had a lot of fun.

The tactical mix ranged from mass rallies to academic reports – and continued to follow the FoE formula of promoting solutions as well as exposing problems. Campaigns on energy issues won public support, as did the group's forays into a closely related subject area, transport. In the late 1970s and early 1980s, the group also launched campaigns on the countryside, acid rain and tropical rainforests.

Each was an intricate topic that couldn't be tackled by guesswork and intuition. Solutions lay in the redesigning of whole sectors rather than tackling symptoms one by one, an approach that had marked FoE out from its earliest days. But for Tom Burke, the challenge was to convert this plethora of top-notch research into dramatic topics the general public could relate to and the local groups network could act and build on.

For instance, in 1974 FoE had published *Wheels Within Wheels*, written by Mick Hamer, an analysis of transport policy and the roads lobby which had a major impact on the philosophy of emerging transport campaign groups. *Getting Nowhere Fast*, also by Mick Hamer, followed. A penetrating analysis of the role of mobility in society, it was so far ahead of its time that it failed to make the impact it deserved.

The campaign to popularise the complex messages these books contained included stunts like delivering a gift-wrapped bike to Edward Heath at Downing Street after traffic snarl-ups just short of total gridlock had stopped him attending Prime Minister's Question Time. Fun and an emphasis on foxy ways to beat the system were still among FoE's strong suits.

The increasing assurance with which FoE used research data to impress and interest the media led some observers to conclude that the group's very nature was changing. The wised-up citizens' group with a big heart was becoming a more elitist, cerebral concern, bent on 'the appliance of science'. In truth, no change was taking place. As always, FoE was using research as part of a tactical pick-and-mix. And in any case, none of the research was purely 'scientific'.

Few were aware of it but science itself was changing in its response to the challenges of understanding and safeguarding the environment. From being pretty much the poodles of industry and governments during and just after the war, scientists were becoming as often as not the whistle-blowers who raised environmental concerns in the first place. This was especially true in the life sciences and their systems analysis offshoot, ecology. Works like *The Tropical Rain Forest* by P. W. Richards, published in 1974, mixed academic with tentatively polemical observations about global forest loss.

Even 'hard science' mavericks like physicist James Lovelock were offering gripping new views on how the planet functioned. Lovelock's *Gaia: A New Look at Life on Earth*, published in 1979, claimed that it lay within human industry's scope to work fundamental changes on global life-support systems such as climate and atmosphere. But the planet could cope. A giant self-regulating system of feedback mechanisms, personified by Lovelock as a super-organism named Gaia, would kick in. Gaia would survive; whether we did was not her problem but ours.

Doomwatch of a more Old Testament type was not going out of business, however. Improvements in the range and detail of available data, combined with refined modelling concepts and great strides forward in the power of available computing systems, was giving rise to increasingly accurate-looking – and increasingly alarming – forecasts of environmental degradation, pollution, resource depletion and overpopulation. The *Global 2000 Report to the President* commissioned by Jimmy Carter in 1978 was a prototype, though the President who would actually receive it, Ronald Reagan, would never read it.

The big question was how development could deliver extra income and resources to support the world's skyrocketing numbers of desperately poor people, yet avoid environmental nemesis. Once seen as mainly a problem of industrial regions, environmental degradation had become what *Our Common Future* would later call 'a survival issue for developing nations, part of a downward spiral of linked ecological and economic decline in which many of the poorest nations are trapped'.

No matter how you dealt the cards, poverty was the ace that trumped every technological fix or development assistance measure. It was ruining peoples' lives, livelihoods and surroundings in one, as major aid charities like Oxfam had begun to realise. Like industrial pollution in the Rich World, if it was left unresolved it could blight the entire planet within decades.

FoE was responsive to poverty and population issues. But the idea of curbing environmental degradation by curbing poverty, and vice versa, in the Rich and Poor Worlds alike, would initially take shape in 1980 in the grandly titled *World Conservation Strategy*, launched by IUCN, WWF and UNEP. The ideas would later be refined – and the buzzword 'sustainable development' would enter the language – in the Brundtland Report *Our Common Future* in 1987.

The *World Conservation Strategy*, which condensed the views of over 400 conservation groups and agencies, was 'wise use' writ large. It shifted the emphasis of debate on living world issues away from rescuing particular wild species or places (and still further from humane concerns over animal welfare) towards a systems view of human relations with nature. An irreducible fraction of the world's wild places should be maintained under protection as storehouses or 'gene banks' where unique and potentially useful species could thrive. Outside protected areas, the natural resource base should be husbanded and harmful new agricultural or industrial developments vetoed.

None of these improvements could be contemplated without a profusion of scientific projects to evaluate them in advance, see them through and monitor their environmental impact after completion. With its insistence on rigorous fact-finding and good science, FoE was merely ensuring it retained a role as a credible voice on the issues of the day. And, at the same time, through national rallies and its local groups network, FoE was still putting people on the street in large numbers. Greenpeace, which would eventually steal much of the media limelight with its aquatic escapades, was not set up in the UK until 1977 and took a while to establish a presence.

FoE was still scoring valuable campaign hits, both nationally and locally. On the energy front, in 1974 FoE Durham had pioneered the use of job-creation funding offered by the Manpower Services Commission to hire unemployed school-leavers and put them to work insulating the homes of pensioners and others living in substandard dwellings. Within four years, over thirty groups were running similar schemes. Ministers in Commons debates were soon commending the groups' contribution to energy conservation – and recycling – as an inspirational example to the nation.

Poland Street and ERR also butted into those debates with publications by Czech Conroy (*Rethink Electric*) and Mike Flood (*Torness: Keep it Green*) which drew attention to existing surpluses of power-generating capacity and pointed out that far bigger future energy gains could be achieved at modest cost through better energy conservation in homes and workplaces than by commissioning yet another batch of nuclear power schemes. The point of these messages was underscored by the Three Mile Island PWR crash in 1979, which missed meltdown and nuclear cataclysm by half an hour and by sheer luck. Within days, FoE local groups organised demonstrations at every nuclear power station in the country.

'Mixing it' on the local and national stage with such issues made sense, when it worked, of the distributed structure of FoE. Local activists piloted user-friendly actions that bore out priorities defined by the national campaigners. Poland Street made sure the issues got noticed in the national media and in the corridors of power.

By these means, the organisation switched the focus of its wildlife campaigning from exotic cats and deep-sea marine mammals to issues closer to home. A campaign led by Poole FoE with Angela King succeeded in getting a scarce occupant of the British countryside, the otter, protected by law from hunting and harassment in England and Wales. Otter hunters in Scotland fought off a similar ban for three more years.

In 1979, FoE marshalled a coalition of cycling and transport campaign groups to mount a 6,000-strong bike rally in Trafalgar Square that went on to 'Reclaim the Roads' by filling Whitehall with cyclists for half an hour, to the blind fury of motorists who had assumed Westminster was all theirs. Earlier that year, cycle campaigner Mike Hudson published a *Bicycle Planning Book*. Intended as a campaign manual for local groups, this was snapped up eagerly by local authority planners at a loss to know how to cater to a sharply increasing use of bikes, noticeable since 1975.

The focus on bikes was also a way to draw attention to the much bigger issue flagged in *Wheels Within Wheels*, namely that the absence of an integrated national transport policy was offering *carte blanche* to road cartels to override the democratic process and urge the Government to implement schemes like the proposed M16 London orbital super-highway.

Figure 12 'Critical Mass': a 6,000-strong contingent of cyclists occupies Trafalgar Square before holding up Whitehall traffic, 1979 © Jan Baldwin

No amount of street credibility was, however, going to be enough to paper over the cracks starting to appear in the organisation's façade after its bruising over Windscale.

By 1978, the growing workload and staff numbers caused strains on space, fund-raising, and on a management structure which had evolved three years earlier, when the staff was half the size. Most of the staff body were responsible to Burke, creating a heavy management load which he was less and less able to fulfil as he was required to spend more and more time at meetings outside Poland Street. The staff had no formal way of voicing their concerns, which were left to simmer.

At the same time, the Board had concerns about FoE's policies and direction, and even about the competence of some staff. The Board was also feeling pressure from the local groups, who wanted more say in the management of the organisation and its campaign strategies. The three-way struggle, with staff, Board and local groups all pulling against each other, led to an intolerable situation and something had to give.

In April 1979 Richard Sandbrook met Tom Burke at the request of the FoE Board of Directors and reminded him of the three-year rule which, as a representative of the original King Street junta, he felt bound to say was overripe for implementation in Burke's case. Burke was angry and shocked to be told it was time to go but he faced a united Board, who offered him the compromise of a long period of notice while a successor was chosen. Following the succession, he would be offered a senior executive post as Vice-Chairman of the Board. Burke accepted and stayed on till March 1980.

But it was an awkward interlude for him, made no easier by leaked stories in fringe magazines and scandal sheets. A report in *Undercurrents* claimed that Burke was 'not resigning voluntarily but being gently but firmly eased out by the Board'. All Tom Burke has to say about the episode now is: 'Richard had a view. I had a different view.' Richard Sandbrook is laconic about it, too:

> Tom had to deal with a very difficult transition, from what had started out essentially as a club, whose structure was cobbled together by chance, to a national outfit with 18,000 supporters, 150 local groups and annual budgets of £300,000. That was part of the tension. The other part was Tom's strong style of management.

After Burke had served out his notice Czech Conroy took over, in the newly created post of Campaigns Director. Burke, in his new administrative post, was set the long overdue task of preparing proposals for a restructuring of the finances and administration of FoE. This was no easy task. The staff were committed to decision-making by consensus and the egalitarian pay parity system. Burke, and some Board members, were keen to abolish these ideals. And the fact that staff were paid at all was sometimes something of a miracle – as the organisation lacked fund-raising expertise money was always a worry.

Clamour within the network for greater empowerment had been growing steadily. A stormy Local Groups Conference had already rejected a token Local Groups Council set up in 1978, as a travesty designed to keep them quiet rather than deal them in. What many wanted instead was nothing less than an elected Board of Directors appointed to govern the national organisation, with members nominated by the local groups.

These concerns came to a head in the second half of 1980. At the Local Groups Conference in October the resented Council proposal was once again on the table. During the fruitless discussions, several local representatives, led by Martin Price of Poole FoE, left the room in anger. This was exaggerated in the reporting to a 'mass walkout'.

A month earlier, Burke had presented his restructuring proposals. All they offered was a new, compartmentalised management strategy that split the top-heavy load of duties presently assigned to the Director into campaigning, administration and publicity portfolios, adding on a new fund-raiser post. Their implementation would almost certainly entail redundancies and the ending of pay parity.

Fearful that the proposed redundancies might affect them and increasingly wary of the Board, the staff formed a union branch within the organisation and informed the Board that they rejected Burke's plans. They were growing bitter, too, about the bad press that FoE was getting. An article by Chris Rose and Charlie Pye-Smith in *Vole* had recently accused them of becoming 'pinstripe protesters' who had lost their radical edge and were now happily tolerated or ignored by the Government. The rebuke cut them to the quick but it held a basic truth.

The deadlock had to be broken. FoE was becoming increasingly paralysed and ineffective. Eventually, in March 1981, a compromise was reached: pay parity would go and some redundancies be made. In return, several members of the Board, including Burke and Conroy, who backed his plans, were forced to resign.

Inevitably, others have a different version of events. Burke recalls, 'We resigned because short of taking over the day-by-day running of the group it would be too much to do to put things right. It was foolish for us to piss against the wind. Then the Chairman of the Board resigned, too.' Burke still feels FoE never recovered from this capitulation which left its structure, he believes, profoundly weakened.

Within a few weeks a Board member, Steve Chipperfield, stepped in as a caretaker Director for a few months until a new person was recruited. In 1981 the top job passed to a former Labour Party manager and Shelter campaigner, Steve Billcliffe.

In October 1981, the air was largely cleared at the Annual Local Groups Conference in Birmingham. No consensus could be reached on the first day of the conference over what constitutional terms the groups wanted. But on the Sunday morning Martin Price canvassed staff and Board members and achieved the impossible – a way forward that all could accept. At the first session that morning he stood up and presented a solution. It worked. A working group of representatives from the local groups, elected by region, was set up to draw up a set of proposals and report them to the Board. The proposals were ratified at a special Local Groups Conference in London four months later.

The new agreement was for a Board with a majority of members elected by the local groups. The new Board members would accept the six Board members who remained after the recent resignations. Other members with expertise in such areas as fund-raising or publicity would be co-opted. There would be no staff representation on the Board but it, in turn, would not interfere with the day-to-day running of campaigns. Local groups also undertook to sign licence agreements, giving the national organisation a little more control than previously over their activities.

One of the co-optees was former Shelter Director Des Wilson, now flagging a new issue, lead-based additives in petrol, through his highly visible CLEAR campaign. By late 1982, Wilson would be chairing the Board and leading efforts to rescue the group's finances. He was to prove one of the most significant players in FoE's history to date.

Wilson was given the powers he needed to bring about the changes he deemed necessary to save FoE from the disaster to which it was headed. The salary budget needed to be slashed – and redundancies were forced through. Pay parity went. Campaigns needed to step up a gear to raise profile and revive morale.

As FoE looked back on its tenth anniversary year it more than likely marvelled at having arrived there at all. Perhaps it took some consolation from the idea that matters were worse outside. Race and poverty riots flared up around the UK throughout the summer of 1981, first at Toxteth in Liverpool then spreading to Hull, Birmingham, Reading, Chester, London's Brixton and thirty other inner-city areas.

Far from enquiring into the possibility that something might be amiss in the quality of the lives and surroundings of the rioting communities, Mrs Thatcher blamed the manipulative involvement of radical troublemakers, 'the enemy within', a category in which she plainly included FoE and other groups interested in boosting environmental quality and the quality of life in Britain's cities and countryside. The Thatcher era was under way, and campaigners used to their old status as darlings of the media had to learn hard facts about making news relevant to people's concerns.

Better times lay ahead for FoE and the green movement during the 1980s. Subscribing supporters of FoE would increase from 10,000 in 1979 through 18,000 in 1981 to 230,000 in 1993. Most of the organisation's peer groups would enjoy a similar boom. But for now, plain and simple survival was achievement enough.

Looking back, it seems reasonable to enquire whether the wrangles over pay parity or the three-year rule were worth all the trouble, seeing that neither of these shibboleths survived in the long run. Cynics might even wonder whether it might have suited some headquarters staff to have the local groups rise up in arms, so that they could resume the more bohemian campaigning approach many of them hankered after, or simply keep their jobs. Maybe it suited some members of the Board equally well, as a way to break the deadlock over the control of management and finances. It certainly looks as though all parties 'played both ends against the centre' to some extent. In the end the deadlock duly broke.

Perhaps, on the other hand, something more fundamental was afoot. The organisation and the movement it spearheaded had been identifiably the creation of middle class, white, principally male and college-educated kids. Could what was happening now be the start of a seismic expansion of that cosily like-minded interest group and its constituency?

In just ten years, the founders and their immediate successors had established in audacious and memorable style the concept that the Earth stood in need of Friends. But could choosing those Friends and championing their hopes and fears for the Earth remain the prerogative of a tiny band of troublemakers?

10

Lie of the land

> The animals sat in the Ark and cried,
> Their tears in torrents flowed
> Who was it said, 'There's land ahead'?
> Encouraging Mr Toad!
>
> – Kenneth Grahame, *The Wind in the Willows*, 1908

Starting in the late 1960s with his work at Shelter, New Zealander Des Wilson made a fine art of the high-profile, single-issue ethical campaign in Britain, deploying it first against homelessness with Shelter then with the Campaign for Lead-Free Air (CLEAR) against lead-based additives in motor fuels. The CLEAR protest hinged on leaked official reports that flagged a possible connection between lead-based pollutants in road traffic exhausts and a rising incidence of certain forms of brain damage, especially among children.

Wilson was first alerted to the lead issue by Godfrey Bradman, a millionaire property developer who was deeply concerned about pollution and had supported anti-nuclear campaigns for many years. He was impressed by Wilson's campaigning talent and agreed to fund CLEAR.

Wilson splashed the cause nationally. The Government was playing down the health link but he sensed a cover-up and suspected politicians of siding with vested interests in trade and industry who saw a ban on lead in petrol as the thin end of the wedge of further regulation. The CLEAR protest soon became a pet cause among the British media and public, building up a formidable head of political steam.

Wilson had long pursued a personal crusade against old-school-tie cronyism and blind-eyeing among top mandarins in Government and big business in his adopted country. Soon after joining Friends of the Earth he also launched, with others, a Campaign for Freedom of Information.

It may be that a generalised liberal reforming zeal rather than a specific affinity for environmental ideas drew Wilson to the organisation. The lousy deal it had got at Windscale was his kind of cause. But whatever his angle, the presence of this self-assured and charismatic mover and shaker put new heart into an outfit still feeling badly bruised and wondering where to go in the wake of the Parker Report debacle. Better still, he arrived with a philanthropic property millionaire in tow. Godfrey Bradman showed willing to underwrite particular FoE campaigns – most notably against nuclear power – if soundly conceived and costed.

Wilson's elevation to the Chair was almost a matter of course. He was soon Campaigns Director in all but name, leaving the administration to a relieved Steve Billcliffe. Wilson was determined to restore the organisation's faith in itself by scoring some major campaign successes. He realised, however, that the bravura single-issue campaigning style he had perfected didn't fit naturally into a broad-spectrum pro-environment enterprise like FoE, nor would it offer the extra involvement local groups legitimately craved.

FoE was on the lookout instead for what strategists would now call 'cross-issue campaigns', clusters of topical concerns that could be lumped together as parts of a themed and measured strategy for eliciting policy reform at the top, or split into single-issue stories or actions that could be used to attract media attention, influence public opinion and keep the funds and support flowing in.

Among the consequences of this quest in the early 1980s would be national campaigns on the Wildlife and Countryside Act, clustered with acid rain and pesticide issues, then on tropical forests in the mid-1980s, linked to consumerism, global trade and Third World development. Transport and energy would figure increasingly among these keynote themes later on.

Wilson's breezy confidence and campaigning track record, along with Bradman's money and the sympathy both men showed for FoE's mission, won them the (sometimes grudging) support of all, even from the more uppity local group representatives on the new Board.

Iris Webb, a science lecturer elected to the reconstituted Board as representative for East Anglia was one of the people who had helped spell out the new deal required to take on board the aspirations of activists working in the local groups. In this process lecturer Martin Price, now the representative for the South-West, also played a prominent part, drafting Rules of Election to the Board and setting up a Strategy sub-Committee to channel local group views on policy issues. Webb says of Wilson:

> There were a lot of things I didn't like about Des, and a lot of things I admired. He had such credibility. He got campaigners to listen to him – the first time for a long time that this had happened at FoE. The only thing that would stop Des dead in his tracks was if somebody laughed at him or teased him. Des was an Eighties Man before the eighties even began, acutely image-conscious.

Martin Price believes that Wilson was personally responsible for FoE's dramatic and happy reversal of fortune in the 1980s.

> Des took the organisation from one that was losing money to one that was financially sound, expanding rapidly and had a clear sense of direction. He

> brought in good staff and trustees. And he handed FoE over to Jonathon Porritt in good shape. Des isn't an environmentalist but he's a damn good campaigner. He gave us a clear political perspective on the issues we worked on and an understanding of what we needed to do.

Webb wasn't the only one who found Wilson's bravura style a challenge to adapt to. National staff working at the new premises FoE had acquired at 377 City Road, between King's Cross and the City of London, prided themselves on their energy-efficient, environmentally sound workplace, designed to house green advocates in exemplary style. Office materials were dutifully recycled and everyone biked to work.

Wilson's daily visitations were seen by some as a blot on this pristine landscape. According to Agriculture Campaigner Chris Rose: 'He used to arrive every morning for a quick burst round the office then disappear in a shiny car leaving a plume of cigar smoke in his wake. All the purists would get upset to see a car parked outside.'

Even Wilson's most outspoken critics concede, however, that his rescue operation saved the organisation from almost certain collapse, a view fervently echoed by friends like Jonathon Porritt.

> You've got to hand it to Des, who came along and said: This organisation has the potential to be what it's always aspired to be and we've just got to turn it around again. And he got money and did the deed. I have to say I think the organisation would have gone under if he hadn't.

Aside from internal developments, FoE staff and volunteers were still campaigning at a feverish pace and nuclear power remained high up the agenda. Inspired by Czech Conroy, the organisation had committed itself to opposing the proposed PWR programme and by the start of the 1980s had already held two huge public demonstrations in Trafalgar Square. The campaign was forecast to last ten years, a prediction which was to prove astonishingly accurate.

The PWR programme was originally intended to consist of a 'family' of six stations around the country, with more to follow. The location of the first was announced in 1982: Sizewell in Suffolk. It was obvious that any attempt to halt the programme must focus on stopping the first and the idea was put forward that FoE should fight the public inquiry.

However, the proposal that the organisation should throw itself into another public inquiry was not met with universal acclaim. There were still many on the staff who remembered the trauma and cost of the Windscale inquiry and the shabby treatment of the anti-nuclear case in the Parker Report. The arguments raged for weeks with the energy team and many local groups sensing a golden opportunity to bring down the nuclear industry while other, equally impassioned, voices saw only financial ruin and wasted effort.

For Iris Webb, who lived only a few miles from Sizewell, the issues seemed clear:

Figure 13 Des Wilson and Jonathon Porritt at Didcot FoE acid rain demonstration, 1985

> I felt I understood the issues quite well. I'd seen Friends of the Earth operate at Windscale and saw no reason to be scared of public inquiries. So when an inquiry was called over Sizewell I couldn't understand why they were reluctant to fight it. We were told the big thing was the money – inquiries were incredibly expensive to fight.

The East Anglian local groups were determined to persevere, even threatening at one point to form an independent splinter organisation and go it alone. At the second budget meeting under Wilson's chairmanship, and following weeks of careful thought and discussion, Webb recalls that 'everyone decided Sizewell was more important than the rest. Campaigners in charge of different budget lines pledged to sacrifice most of their resources to fight it. But half the Board still said finance was grim, we were going to the wall. We said if we're going to go down, let's go down fighting.'

With the decision made to go ahead, it was vital that the organisation avoid the naive optimism that had been so violently crushed at Windscale. No one on the staff was under any illusions that the inquiry could be 'won' but everyone was clear that they had a perfect opportunity to decisively weaken the PWR programme. After all, they reasoned, if FoE didn't attend there would be no one to make the safety case against PWRs and any future inquiries would go through on the nod.

In the event, the FoE Sizewell team, initially led by Energy Campaigner Renee Chudleigh, achieved far more than they had ever hoped for. The team's Researcher, William Cannell, had put together a brilliant case against the supposed safety of PWRs and an inquiry which was intended to last a year quickly developed into a marathon which only finished in 1985 after twenty-seven gruelling months.

Nor were the staff by any means alone in their work. Local groups throughout the region worked frantically to oppose Sizewell and even Graham Searle appeared at the battlefront once again. Searle – now a resident of Suffolk – attended every day of the inquiry and led the incredibly effective Stop Sizewell B Association. If the Sizewell proposal went through, it would not be for want of opposition.

Finances, however, could not look after themselves and a huge parallel operation was soon underway to raise the desperately needed cash that the inquiry was swallowing. Fortunately, the list of donors had grown exponentially as events at Windscale had unfolded, to around 19,000 by 1980, so there was at least a mailing list to be milked.

Iris Webb, Briony Jones and other national office staff wasted no time in rattling the can and they had raised £43,000 by Christmas 1982, mainly through direct mailshots, pledges from local groups and philanthropic donations from the likes of the trustees of the Benjamin Britten Foundation, based next door to Sizewell. The money was important: FoE's final bill for the Sizewell inquiry would be £250,000, without staff costs.

Meanwhile, back in the inquiry rooms, the energy team were using every trick in the book to make an impact. Cannell's devastating safety

Figure 14 Sizewell PWR inquiry, 1983

case against PWRs was shaking the Government's Nuclear Inspectorate out of its earlier complacency and putting nuclear safety issues on the political map. Nor did the CEGB have it any easier: ultimately they would have to redesign parts of the station as a result of FoE's critique and incur major extra costs.

It was, perhaps, a sign of the organisation's continuing ability to worry the Establishment over nuclear issues when an independent TV documentary called *First Tuesday* revealed that the phones of Iris Webb and other campaigners had been bugged. The UK internal security service, MI5, was taking no chances with the anti-nuclear lobby and had taped campaign meetings and eavesdropped on telephone conversations during the inquiry. Webb says,

> The interviewer for the programme said he expected me to be more shocked than I was by the surveillance evidence. But I expected it. It was an intrusion but it certainly wasn't a surprise. We were disagreeing with the government, Margaret Thatcher called us 'the enemy within'. She'd reached that point where someone who doesn't think or behave like you or your peer group is no longer just someone with a different idea, they're the enemy.

When the Sizewell inquiry finished in early 1985 the energy team were under no illusions that they had 'won'. But they could rest assured that the PWR programme was firmly in the media spotlight and that nuclear safety was at the forefront of the public mind. The final inquiry report itself was due to appear in late 1986. A few months before it was published another nuclear power station was to catch the headlines – Chernobyl.

Stewart Boyle, a former Personnel Officer with Coventry City Council and local campaigner for FoE, had become Energy Campaigner in October 1984, midway through the inquiry. He was eventually to inherit the aftermath of the Sizewell inquiry and much besides. Boyle sees the PWR battle as crucial to the eventual halting of nuclear power:

> We didn't expect to win at Sizewell but knew that we could score a lot of points and slow down the PWR programme. In the event the timing proved crucial; if Friends of the Earth hadn't gone into the Sizewell inquiry it would have been over in a year and the final report would have come out before the Chernobyl disaster.
>
> In the end, the report had to appear after the disaster and into a public climate that had hugely changed and was very anti-nuke. You have to remember that they originally wanted to build six PWRs for starters and then a lot more. In the end they only got Sizewell, and that didn't go on-line until the mid-nineties. Friends of the Earth took a big risk in fighting the inquiry but I guess ultimately fortune favoured the brave.

Sizewell and Chernobyl had not been the only nuclear landmarks of the mid-1980s, however. An equally feverish battle had also been fought across the country on reprocessing – or, to be more accurate, the dumping of reprocessing waste from Sellafield.

A landmark report by William Cannell – the cheerfully entitled *Gravediggers Dilemma* – was released in February 1985 and laid out the intellectual case against nuclear reprocessing in withering detail. The conclusion of the argument was clear: reprocessing was a disaster that would generate 160 times the nuclear waste that would result from not reprocessing.

The figure of 160 was crucial and the campaign team hammered it home at every opportunity. Stewart Boyle, who coordinated the campaign, remembers: 'When that magic figure 160 started turning up in Steve Bell cartoons we knew we had finally planted it in the public mind.' And it was the inexorable logic of the 'magic figure' that would force the nuclear industry into trying to find somewhere to dump the waste they planned to produce.

And the list of dumping options was contracting all the time. Greenpeace had recently won the battle against sea dumping of nuclear waste, which meant that a land site would have to be found. A special body was set up, called NIREX and allegedly separate from BNFL, to hunt for what was rapidly becoming the Holy Grail of the nuclear industry – a new site for Sellafield's radioactive rubbish.

Things were not set to run smoothly, however. In mid-1985, FoE received a leaked document through the post: a highly secret NIREX map of potential dump sites. Generous as ever with their information, the

Figure 15 Chernobyl anniversary demonstration: protesters march to Sizewell A nuclear power station, 1987 © H. Fricker

energy team sent copies of the map to every local authority with a potential site in their area. They also offered more information about nuclear waste and its hazards.

The response was overwhelming. FoE staff and local group activists addressed over eighty separate local authority meetings as well as countless public gatherings. Councils fell over themselves to declare an anti-dump policy and donate funds to FoE's fighting fund. The planners at NIREX, meanwhile, saw an already thankless task becoming ever less welcoming.

Under a political barrage and desperate to make progress, NIREX announced in late 1985 a reduced shortlist of four possible sites. FoE duly ran campaigns in all four areas, pressing MPs, training local groups and whipping up the media. NIREX found itself harried at every meeting and continually battered by hostile public opinion, and six local newspapers ran anti-dump campaigns and gave away 'no to NIREX' posters to their readers. Stewart Boyle remembers all too well the emotions that the campaign stirred up: 'Local people were furious about plans for a dump. NIREX staff were literally run out of town on occasion, refused accommodation in the local pubs and told to push off.'

Finally, the shortlist came down to two sites and national politics intervened. Both the sites were in Conservative-held constituencies open to being overturned by a wave of anti-NIREX emotion. Six weeks before the general election in 1987, both the sites were declared 'unsuitable' and NIREX was sent back to the drawing board.

'The whole campaign was a classic pincer movement,' recalls Boyle. 'Greenpeace shut down the sea disposal route and we hit them on the land. After NIREX had abandoned their shallow dumps they were left with the prospect of a deep burial site and that was going to be hugely expensive.'

And if the anti-nuclear atmosphere of the late 1980s wasn't enough for an embattled nuclear industry, worse news was soon to follow. The Government's commitment to free-market economics was to drive it inexorably towards privatisation of the electricity industry and a sell-off of all its assets, both nuclear and conventional. All of the highly cost-sensitive safety issues that FoE had raised in the 1980s would prove to be much more than simple campaigning points. They would end up as vital elements in the calculations of the City of London's financial markets. Markets that were to prove every bit as unsympathetic to nuclear power as the most ruthless green campaigner.

Aside from the nuclear campaign, and perhaps as a counterbalance to it, other campaign fronts were opened up in the early 1980s. Des Wilson backed campaign staff who wanted urgently to diversify the group's campaign portfolio, to build on its track record on behalf of wild species and places, and make new stands on pollutants and hazardous wastes in industry and agriculture. Godfrey Bradman agreed to bankroll new research by ERR into pesticides (conducted by Maurice Frankel and others) and a probe into the emerging acid rain syndrome by Nigel Dudley and David Baldock.

The initial point of impact for new campaigning was, however, the passing in 1981 of the Wildlife and Countryside Act. This long-awaited piece of legislation was billed by Whitehall press officers as a showcase approach to dealing with environmental issues in the countryside, notably problems that hinged on conflicts between intensive farming and the protection of habitats important to native wildlife.

In the view of FoE, the new Act's provisions were a charade, most of all where SSSIs (Sites of Special Scientific Interest, Britain's most important wildlife sites) were concerned. Some politicians were in the habit of calling these sites Britain's 'national nature reserves'. But in reality they amounted to nothing like a methodically run protected areas system. These jewels in Britain's conservation crown were continually sacrified to destructive development pressures, in particular intensive agriculture, plantation forestry and roadbuilding.

Despite a long agricultural depression that had started in the 1870s and lasted through to World War II, and despite the break-up of many great landed estates after World War I, repeated calls for Government action to create a national network of specially protected areas and guarantee public access to it, had been stifled time after time by vested landowning and farming interests at Westminster. Conservation measures introduced in 1949 under Town and Country Planning legislation had myriad objectives but this was not one of them.

They were primarily aimed at curbing the adverse impacts of industry, urbanisation and leisure on the natural scene. Progressive planners had long argued for curbs on 'ribbon development' of new housing along highways, and on other forms of urban sprawl. They wanted an integrated approach to town and country planning to keep the intrusions of industry and recreation-bent urban refugees within planned limits.

Since 1949, as Bill Adams points out in *Future Nature*: 'In various ways each of these impacts had proved harder to live with than had been expected . . . However, none did as much damage to the countryside as the two industries left uncontrolled by conservation: agriculture and forestry.'

Norman Moore, Chief Scientist of the Nature Conservancy Council and one of Britain's top ecologists, had observed in 1969 that in thickly populated countries like Britain, land that can be set aside for wildlife is necessarily limited, so wildlife has to be protected outside reserves, on land whose primary use is for agriculture, forestry or recreation. On such land, Moore warned, conservation requirements tended to 'clash to a greater or lesser degree with the primary land use'. He pointed to larger farm and field sizes, mechanised farming methods, blanket use of agrochemicals and rising yield targets as evidence that farming was breaking with traditional patterns of land use that tended to blend with conservation.

In 1980 the release of Marion Shoard's boldly incisive book *The Theft of the Countryside* raised the temperature of debate over the impending Act. Shoard bewailed the negative impacts of intensive agriculture on the quality of Britain's rural environment. She sharply questioned the point of the abundant subsidies and tax breaks farmers and landowners enjoyed, born of a fixation successive postwar governments had on achieving national food self-sufficiency in case another war broke out.

The production levels needed to secure this guarantee had long since been surpassed, said Shoard, but land clearing and intensification of farm production continued apace at the expense of the taxpayer and at the cost of an increasingly de-natured rural landscape. The quality of rural community life was suffering, too, as mechanisation drove labour from the land. Rural house prices had skyrocketed and in many places village life had become a vacuum as property speculators encouraged affluent city folk to buy weekend homes in the country.

FoE had been pressing since the beginning for statutory planning controls over agricultural development, especially where it threatened habitats important to wildlife, such as long-established wetlands, woodlands, hedgerows, heaths and uplands. In 1981, FoE launched its Countryside Campaign with the manifesto *Paradise Lost?*, which echoed and extended many of the points raised by Marion Shoard. It also indirectly challenged established conservation groups such as the RSPB and Royal Society for Nature Conservation, to take a firmer stand.

Most of these groups were prepared to go along with a so-called 'betterment' principle recommended by the Nature Conservancy Council, the statutory quango which was the Government's main consultative partner on new countryside legislation. Enlightened self-regulation was the NCC's implicit answer to winning the cooperation of landed interests in moves to safeguard conservation-worthy sites such as designated SSSIs. A wider consensus on maintaining the quality of landscapes and countryside in general could then be built, it was implied, upon this irreducible foundation.

A groundswell of dissent from this view was already evident among fringe groups like the British Association for Nature Conservation, a group started by Bill Adams, Chris Rose and Charlie Pye-Smith. Rose was a graduate of the MSc Conservation Studies course founded in 1969 by Max Nicholson at University College, London. He and his fellow BANC founders typified a new wave of progressive environment specialists coming out of British colleges and universities in the 1970s.

BANC published a journal called *Ecos* and also used the alternative press to voice a more proactive view of topical environmental issues, well-informed on the scientific front but with no fear of courting controversy. In one *Vole* article, for instance, Rose accused Whitehall of neutering the NCC, the Forestry Commission and other statutory bodies that were meant to protect the environment, by stacking their Boards with placemen sympathetic to farming and landowning interests.

A renegade member of the landed gentry, Lord Melchett (the same Peter Melchett who would later run Greenpeace UK) was meanwhile asking FoE, WWF and others to back a new consortium, Wildlife Link, set up to press for more radical statutes for conserving habitats. He invited Rose and Pye-Smith to represent BANC and ginger up what he saw as the Link's stodgier members. FoE leased Melchett part of their City Road premises to use as HQ offices.

FoE was keen to shift groups in Wildlife Link and the NCC towards an irreversible pledge that there would be no further destruction or pilfering of SSSIs. To date, the NCC's response when SSSIs or parts of them were lost to development had been to designate a new site or parcel of land elsewhere, so the total acreage of SSSIs never appeared to vary. When the Wildlife and Countryside Act was passed in the summer of 1981 it made notification of SSSIs the critical core of official nature conservation policy for Britain. But it did nothing to put a stop to such arbitrary gerrymandering.

At about this time, a fortuitous partnership grew up between FoE's Wildlife and Countryside Campaigner, Charles Secrett, and Chris Rose, who was by now a freelance journalist and co-writing a book on countryside issues with Charlie Pye-Smith. Secrett and Rose were outraged by the Government's refusal to safeguard the SSSI network and determined to do something about it. They formed an oddly effective duo, the lanky Rose, nobody's fool and a media expert with a sharp eye for a racy headline, and the more saturnine and strategy-conscious Secrett.

In 1982, FoE published the first results of their joint labours, *Cash or Crisis*, a thorough and damning indictment of the flawed legislative, financial and technical thinking behind the Act. In some ways these frailties were an accident of party politics. The original draft Act, introduced by the previous Labour Government, had placed a legal responsibility on landowners to ensure proper safeguards over SSSIs on their land as a statutory duty of care. But this formulation was short-lived. When the Tories took over they reversed the onus. The taxpayer would have to pay landowners instead, to make protection worth their while. But at that time the Act's sponsors only had in mind a selection of around forty key sites, out of 3,800 then gazetted as candidate SSSIs. From the process of considering more than 2,300 amendments at the Act's passage through Parliament, it was established that *all* such sites ought to be included. This shift raised acute problems of financing. Says Rose,

> We did the sums on the back of an envelope. I think the Treasury quite literally did the same and concluded that this move was a non-starter. They were in a cleft stick because to avoid forcing the hand of the farming lobby, the Ministry of Agriculture, the Forestry Commission and all those big landowners and forestry investors, they were about to let themselves in for what amounted to a giant tax sink.

Anchors were thrown out, as much to avoid imputations of State interventionism on a grand scale as to cut the Act's costs.

Efforts were concentrated on making the Act's provisions work on a so-called voluntary basis rather than on purchasing sites and having them conserved under specialist supervision. Owners of the sites were instead to be offered grants along with powers of stewardship. A farmer could keep, say, an ancient watermeadow on his land just as it was and receive up to £100 a hectare not to do anything to it.

Rose translates this provision as saying,

> well, we're just going to have to turn the farmer into a sort of stand-in gamekeeper or warden. Which farmers were mostly quite happy to do, especially a few years later when they were being given enough money and the price of corn was beginning to drop. They thought they were on to a good thing. I remember one farmer responded to a press article I wrote questioning this arrangement. 'We farmers are all in favour of the Voluntary Approach,' he told me. 'We're in favour of compromise and now this Act gives Government the teeth to enforce this voluntary compromise.' In other words, they got exactly what they wanted plus wads of public money.

The cost of these 'management grants' – in effect compensation to landowners for profits forgone by not developing the sites – topped £10 million in 1982 and grew rapidly. By 1990 it was costing over £38 million. FoE and others argued that the grants were a waste of money. Moreover, they didn't even work: SSSIs were still being harmed and lost at a steady rate, sometimes even prior to baptism.

A 'three-month rule' allowed unscrupulous landowners to develop newly-proposed SSSIs without hindrance during the twelve week period before notification was finalised. This loophole would be closed in 1985 by a second Act. But the notification process continued to prove awkward and costly. According to Charles Secrett,

> What we were challenging were three fundamental assumptions on which the Act was based. The first of these was that the voluntary approach was all you needed, that you could rely on the good instincts of landowners and developers to respect these precious items of natural heritage and not to bulldoze them or plough them up or cut them down. Of course the private sector wasn't going to respond unless restraints like these were backed by law.

The second objection – and the central argument in *Cash or Crisis* – was that the financial inducements in the Act amounted, in Secrett's words,

> to little more than bribes, to compensate landowners for not destroying what had been scientifically determined as our most important wildlife assets. We believed that was entirely the wrong way to go about protecting for the nation what the nation itself had established as its prize heritage. Friends of the Earth argued instead that public money should be used to encourage the right sort of management action by landowners who genuinely needed financial help to fulfil their duty of care. But not to hand it out automatically to people who could well afford to protect that which should be protected anyway.

Grant support should, in other words, be concentrated more on the types of development the voter and taxpayer wanted to see in the countryside. There should be a far greater sense of public involvement in decisions taken in the political arena that would affect the future shape and use of the land. And this more democratic imperative should apply not only to habitat conservation in specially protected areas but also to the farmed landscape and the British countryside as a whole.

The third main strand in the new Wildlife and Countryside Campaign agenda was to advocate a more open theatre of debate, bringing the wishes of ordinary people into the picture alongside the narrow range of landowning and land use interest groups normally consulted on statutory conservation measures. 'So much of the campaign was about waking local communities up to threats to the natural heritage on their doorstep, showing them what they could do to help be part of the solution rather than sitting idly by and letting things happen that they didn't like', Secrett explains.

In tactical terms, these strands came together in the way that FoE began working with local communities to take peaceable direct action – in the early 1980s – to protect threatened SSSIs and other cherished areas. Secrett's proposal to get activists and local people to sit together in front of bulldozers – or whatever was destroying an SSSI – was approved, even though the organisation had little or no experience of this type of campaigning. 'It felt like a step into the unknown,' he explains, 'but it also felt the right thing to do. We couldn't just stand by and let these precious havens be lost.' A number of successful site battles were engaged around the country, in Dorset, Essex, Kent and Halvergate Marshes in Norfolk.

Some were successful, and the sites saved; others were not, and the developers had their way. But all generated considerable media coverage, public debate, and, most importantly, a sense that ordinary people could make a difference. Overall the campaign was a crucial shot in the arm for the countryside conservation movement, helping to galvanise public opinion and turn the destruction of SSSIs into an issue of political importance.

The star of the Halvergate Marshes campaign was an articulate and expert troublemaker, Andrew Lees, who was to play a prominent role in raising the organisation's national and international profile in years to come. For now, local activism in Norfolk, inspired and tirelessly championed by Lees, stood out as a model of participatory neighbourhood campaigning. It showed local communities an effective way to stand up for the land they knew and loved, confident that they were backed up by a national movement mounting similar actions countrywide.

Such actions made ideal use of the organisation's distributed structure and showed the way towards an innovative role for FoE as a pioneer of an interactive grassroots campaigning style that few comparable groups could match. Charles Secrett, for one, sees this aspect of the Wildlife and Countryside campaign as by far its most significant outcome:

> It's often forgotten that some of the main roots of the participatory forms

of environmentalism that Friends of the Earth and others are so set on today can be found in the Countryside campaigning that we geared up in the very early eighties. Campaigning tended to be very much co-ordinated and directed from the centre, building on methods we developed in the seventies. But from when Brower first set it up in America as a counterbalance to the conservative approach taken by the mainstream organisations, Friends of the Earth has always had a devolved network of groups that were virtually autonomous entities, taking their own decisions about priorities. They are people you work with, you don't tell them what to do.

The Wildlife campaign which Secrett inherited in 1980, focusing on whales and endangered species, had encouraged local group activists to show people the power they had as consumers and as voters to change things politically or drive markets and consumer demand away from potentially destructive pathways. 'I was building on that tradition, showing it wasn't just about what was happening globally: everyone had opportunities to play a citizen's role in shaping events in their backyards,' he says. 'Opportunities grew for local groups to take forward their own campaigns in their own way, focused not only on official SSSIs but also on locally valued sites that weren't designated. The groups had always done this, of course. We started to support them in a more coherent way.'

This wider compass also began to reflect in local responses to the agriculture dimension of the campaign, to national and international issues like acid rain and the health hazards of pesticide sprays, as well as to battles over local protected areas, amenity sites or places of recreation.

Chris Rose, who joined FoE in 1983 and whose work on acid rain and pesticides issues had as its main target the influencing of policy change at national level, none the less found himself working more and more with and through local groups to stir up interest and controversy around the country and to seek advice. He says they were a key factor in national campaigns on agriculture issues.

Charles and I were hammering away. I'd started to do a lot of work around taxation fiddles involved in forestry. But we were getting frustrated because when we complained about fields of rare wildflowers being ploughed under or 800-year-old hedges being blown up with dynamite, you'd still get the press talking in terms of environmentalists standing in the way of progress, saying things like: Don't you want to eat? Isn't it the inevitable price of progress? Even *Guardian* reporters and such, liberal types who'd now be outraged, were saying it.

So we thought, we have to have something that torpedoes this impregnable bucolic *Archers* image of agriculture and farmers. Because till we do that, these finer arguments about land use or management will all be academic, no-one will listen. So we needed to find something a hundred per cent unacceptable to the public that was to do with agriculture.

Rose consulted with local groups, including a group based in Esher, Surrey, that included several farming people. They narrowed the range of potential anti-Archer issues down to two that hinged on agricultural wastes; the post-harvest burning of straw and stubble in fields, and the disposal of slurry (liquefied manure) in streams and rivers. The local group activists told Rose they'd already tried slurry to no effect, so straw-burning was decided on as a showcase issue.

It proved a timely choice of campaign priority, coinciding with one of the hottest and driest summers on record. A stark choice was graphically posed between clean air and 'progress'. Rose recalls,

> There were enormous fires running out of control all over the country. Hospitals had to be shut because ventilators couldn't work in operating theatres, people died in car accidents caused by the smoke. There was an absolute national uproar for two or three months. Farmers were completely split on what to do about it. It had quite an effect on puncturing their reputation.

It also led to immediate local curbs on straw-burning the following year and – in due course – a total national ban on the practice.

Campaign principals felt keen to follow up the moral edge this breakthrough had given the organisation. Having recently taken delivery of the research on pesticides that Des Wilson had commissioned from ERR, they felt it was time to extend the organisation's track record on Food and Environment questions into a nationwide campaign about pesticide use and regulation, to go public in early 1983. Shortly after, as research into acid rain also came on stream, industrial pollution issues would merge with agriculture, energy and transport issues.

The Science First emphasis that had been a main feature of FoE from the start seemed set to remain the mainspring of its work in the 1980s. Yet the new campaigns were to prove that the 'appliance of science' had shifted ground in subtle yet profound ways since the countryside had come to town.

11

Seeing the wood

> The belief that science proceeds from observation to theory is still so widely and so firmly held that my denial of it is often met with incredulity. I am even . . . accused of denying what nobody in his senses can doubt. But in fact the belief that we can start with pure observations alone, without anything in the nature of a theory, is absurd. Observation is always selective. It needs a chosen object, a definite task, an interest, a point of view, a problem. And its description presupposes a descriptive language, with property words; it presupposes interests, points of view and problems.
>
> – Sir Karl Popper, *Conjectures and Refutations*, 1953

In a twenty-years-on sequel to *Limits to Growth* called *The First Global Revolution*, the Club of Rome declared that the complexion of environmental issues had changed almost beyond recognition during the 1980s. Formerly, the report claimed, most kinds of environmental deterioration were 'essentially local and could be eliminated by local and national action, at a cost certainly, but one which could be borne'. Now, however, things were different. 'Environmental threats of a new order of magnitude and difficulty have been identified, which demand quite a different approach. These reside in a number of macro-pollution phenomena which are global in scope and beyond the capacity of individual countries to eliminate.' The report went on to identify these new, costlier threats as the accumulation and trans-boundary transfer of toxic and hazardous wastes from industry, agriculture and nuclear installations, the acidification phenomenon, the depletion of the ozone layer and the 'global warming' effect of surplus carbon dioxide and other 'greenhouse gases' in the atmosphere.

The authors of the report formed the Club of Rome Council, an *ad hoc* consortium of senior worthies that included lawyers, economists, scientists, philosophers and policy pundits from around the world. But this group was plainly a different creature from the proto-Green *Limits to Growth* team.

It showed its colours in self-important observations like: 'The green movement, useful as it is, may be inadvertently diverting public attention from these longer-term and more serious environmental issues by impressing the person in the street with easily appreciated, immediately visible but strictly local damage.' Implicit in this put-down was the view that the new global problems were too complex to entrust to groups like Friends of the Earth, or to anyone who did not dine at the top table of science, policy-making or lawgiving. Only anointed experts could be trusted to make judgements that were 'value-free' as distinct from rabble-rousing or emotive.

This lofty analysis reflected a widespread upgrading of professional concern for the environment in governmental, legislative, business and academic circles during the 1980s. The Club only faintly praised grass-roots environmentalists for putting most of the key issues on the map in the first place and overlooked the preposterous dances officialdom had often led them. It also appeared to dismiss any idea that 'persons in the street' might be intelligent beings able to appreciate issues put to them without guile or hokum, and make personal or communal decisions about how to respond.

The report's most profound flaw was the idea it implicitly promoted that reconciling environment and development was a deed that could only be effectively performed from the top down. Lawyers, economists, policy makers and scientists might design solutions to destructive development that made perfect sense to lawyers, economists, policy makers and scientists. But lawyers, economists, policy makers and scientists do not 'do' development: farmers, factory managers, fishermen, miners, foresters, builders and consumers do that. Their actions are in the end the actions that make a difference.

A fair case could be made for reinventing the apparatus of law and policy so that it did not pull wool over the eyes of the 'person in the street', actively encourage environmental abuse or deliberately impede social changes leading from the bottom up to improved environmental prospects. And the Club was right to imply that there were drawbacks to doomwatch scaremongering over environmental issues, if unsupported by at least enough hard evidence to cast reasonable doubt on the comforting notion that the planet's fate was in safe hands.

But who, in the end, was best qualified to judge what was or wasn't 'reasonable' doubt? The scientist? The politician? The industrialist? The media? The UN? They all had their share of skeletons in closets. Even the much-vaunted objectivity of the scientist had been debunked by top science guns like Karl Popper, as a counter-productive myth too often used to bolster entrenched power structures. Whose environment was it, anyway?

If pro-environment protest drew the regular citizen's notice to 'easily appreciated, immediately visible damage', why should that distract attention from the bigger issues? Today's protest was the raw material of tomorrow's orthodox morality; its functions included a mission to challenge official complacency by flagging any evidence of threatening trends well ahead of the authorised version of history.

As environmental pressure groups matured, they learned to employ scientific data and standards of evidence as a safeguard over credibility. What they sometimes seemed in danger of forgetting was that they, too, were capable of putting science on a pedestal to which it had no claim. Science deals in verifiable evidence but scientists hardly ever agree on the interpretation of that evidence, let alone use it to contradict the view of their clients and paymasters whether these happen to be Prime Ministers or pressure groups.

FoE had begun learning this lesson the hard way at Windscale and by 1983 had learned enough to count to ten before using scientific expertise as an opening gambit. Says Chris Rose,

> We had tomes of pesticide research, lists of banned chemicals, data about toxicity and so forth. To put them to use, we intended to campaign in the first place over whether people should have freedom of information on this stuff. Then to our surprise, we started getting phone calls from members of the public saying they'd been sprayed.

What ensued was an unreservedly populist campaign. Casting caution to the winds, FoE started publicising first-person accounts by people suffering chronic health problems that they felt sure dated from exposure to pesticide sprays. Through the media, other people who had similar incidents and problems to report were urged to get in touch.

There was a flood of responses. It turned out that in many cases protests had been lodged with the Health and Safety Executive but had been ignored. A dig through the Public Records Office revealed that doctors and medical officers had lobbied in the 1950s to have hazard signs placed in every field that was being sprayed, warning people to keep out. Rose says,

> It was all basically anecdotal, but our argument was: if all these people feel sick or think they're sick because they've been sprayed then there's some problem here. Especially if no medical tests were being performed to prove no poisoning had occurred. Some people said they'd had abortions on the advice of doctors after inhaling pesticide sprays but the doctors had refused to confirm this after.

A number of MPs, led by Richard Body, took up the question in Parliament. Body was on the Agriculture Select Committee and filed a Minority Report on Pesticides that led to an official inquiry. Various local Public Health Officers gave evidence of health hazards from sprays and hospital surgeons affirmed that people exposed to sprays were prone to added risk of cancer.

'In the middle of the campaign, when Government spokesmen were busy playing down these revelations and insisting that there was no problem, the then Treasury minister Leon Brittan piped up in Cabinet to say: yes there is! Some bloody helicopter sprayed me yesterday in my back garden!' Rose recounts. After this embarrassing contretemps leaked out, the campaign really began to resonate with the public. 'They didn't like this idea of the countryside being one great big spray zone.'

As well as further blotting the Ambridge copybook, the campaign enraged the wealthy and powerful agribusiness lobby, which was already under heavy pressure from other quarters. Oxfam's David Bull had recently published *A Growing Problem: Pesticides and the Third World Poor* which revealed that there were some 11,000 deaths a year by pesticide poisoning in developing regions and many more unrecorded cases of illness.

Some pesticide compounds that were banned from use in industrialised countries were still being marketed, manufactured and used in parts of Africa, Asia and Latin America, often without adequate safeguards in field or factory. A tragic confirmation of these warnings would come in 1984, when a leak of methyl isocyanate from a Union Carbide pesticide factory in Bhopal, India, killed 3,000–8,000 people and injured 200,000–600,000 more.

The environmental as well the public health costs of intensive pesticide use were also under scrutiny in the UK and Europe. The ill-effects of pesticide use and abuse on the ecology and water quality of streams, rivers and wetlands downstream from farms and pesticide factories were becoming matters of headline concern. Still on the rampage in Norfolk, Andrew Lees blew the whistle on a chronic mercury pollution problem that was poisoning fish and other aquatic life in a long stretch of the River Yare. In a clever piece of detective work, he had a number of eels from the river analysed and proved conclusively that the level of contamination was illegal and dangerously high. It was alleged the mercury was coming from the May & Baker agrochemicals plant near Norwich. Following a vigorous local campaign, which also featured in the national press, the discharge was stopped and the river began a slow recovery.

A few years later, in 1986, 30 tonnes of toxic waste was discharged into the Rhine from a Sandoz pesticide plant near Basel, Switzerland. Though eclipsed by a far more dramatic environmental catastrophe in the same year, the Chernobyl nuclear disaster, this incident helped speed a flood of water-quality regulations and directives through the channels of the European Commission, along with added regional curbs on the use and disposal of chemicals used in industry and farming.

Greenpeace was tackling pesticide issues and related pollution problems from a different angle, embarking on an astute bioregional campaign to force the governments of countries bordering the North Sea to clean up their regulatory act in the sea's margins and feeder rivers. As well as mastering an intricate array of 'soft law' levers to shift this agenda, Greenpeace was also busy taking on board an increasingly slick research capacity during the 1980s, with a view to proving that industry did not have a monopoly on technical wisdom.

Arch-doomwatcher Barry Commoner had warned in the 1960s of a 'closing circle' of toxic chemicals in the environment. Was this circle at last showing signs of loosening its grip on Europe? At least it looked as if its measure was being taken. Some enthusiasts for Britain's environment began to feel optimistic as manufacturers and major users of toxic chemicals were forced more and more onto the defensive.

But any such optimism was soon stopped dead in its tracks by the official response to growing disquiet over acid rain pollution and other ills linked to the burning of fossil-based fuels. In the course of its campaign on these issues, FoE would find itself in head-to-head confrontation with its old adversary at Windscale, the Central Electricity Generating Board (CEGB) and with a monetarist policy doctrine at high noon under a still 'un-greened' Margaret Thatcher.

Though terms like acidification or acid deposition best define the acid rain problem (which takes many forms besides rain), its main vehicle is rainwater tainted by sulphur dioxide emissions from factories or power stations where coal and other fossil fuels are burned. Reacting in the clouds to form dilute sulphuric acid, these exhausts can poison freshwater life far from their source. By the time the problem began to be widely recognised and debated in the late 1970s, most fish and other aquatic life had been wiped out in the lakes and rivers of southern Norway and western Sweden. Scientists there were convinced the problem was imported on the wind, from Britain.

Around the same time, signs of acid overload began appearing on land, too. Millions of conifer trees began to wilt, turn brown and die back in woods and forests all over northern and central Europe, especially in Germany. Acid rain soon became the top environmental issue on the European mainland. Most industrial nations acknowledged the threat and agreed in 1983 to curb emissions from these sources by 30 per cent within a decade.

Another, more pervasive, acid threat was emerging too – waste oxides of nitrogen (NOx for short) issuing from motor exhausts and from countless other domestic and industrial sources. Their impact on life and landscape seemed to be much like that of sulphur-derived pollutants, though they acted and spread more locally at first. Rising background levels of hydrated nitrogen (ammonia) and nitrates from farm wastes were also identified as a factor, triggering acid effects when they accumulated to saturation point in exposed soils and waters.

In the teeth of a mass of evidence from the continent, the British Government took the official position that no acidification of any kind could be positively linked to pollution problems. The effects could, they argued, just as easily be driven by natural processes or local land use changes. The scientific civil servants of the Forestry Commission insisted that Britain's forests were safe from harm. The CEGB was awarded a £5m Government grant to launch long-term scientific investigations into the underlying causes of acidification, effectively stalling the issue. Chris Rose recalls,

> With acid rain we were basically starting from zero. I was given these piles of research and had to go away over Christmas and convert myself from a countryside campaigner into somebody campaigning on acid rain. The management said, look, money's short and we can't have two countryside campaigners. You're a scientist, you'll understand all this stuff, go and learn it.

At first, FoE saw no reason to doubt the Forestry Commission's assurances that the forest dieback problem was limited to the continent. 'Then I went to the Netherlands and looked with foresters and Dutch ecologists at their trees,' says Rose. 'They showed me different symptomatic changes and side-effects caused by acidification. Later on I went to the Black Forest and saw more of them. I realised, this is just like the trees in England, it's the same thing.'

A long battle ensued, causing much media stir. Mark Neville joined with Rose in efforts to persuade officialdom that power station emissions and other fossil fuel exhausts were not only wrecking other countries' natural resources but also blighting Britain and threatening its rural economy. FoE invited top foresters from the Netherlands and Germany to come to Britain and confirm that the signs were all there. 'Gradually the experts came round,' Rose relates, 'though they still haven't come right round even today. First it was the scientists, then the private foresters. After a few years they all said, yes, it's happening.'

In National Trust parklands around the country, famous ornamental trees were dying back.

> Tree surgeons kept trimming bits off their Cedars of Lebanon and such but the trees kept dying and dying. We said there must be something going on, it's not drought or some other natural effect doing it.
>
> We were aware it could take a hundred years of research to prove it. Studies on acidified lakes, backed by the Nature Conservancy Council, had knocked the CEGB's arguments about natural causes on the head. Data from the forest were much harder to interpret, though, harder to get at, more complicated. It meant dissecting what was happening with forest trees, soil fungi and soil changes, dry fallout, wet fallout, ambient pollutants, secondary effects, primary effects and so on.

The Government-backed CEGB research, steered by the Forestry Commission, zeroed in on soil effects and took a reductionist science approach to the issue. Says Rose,

> They insisted on replicating everything everybody else had already done, only much bigger because they had more money. All this shadowing actually drove quite a lot of Institute of Terrestrial Ecology scientists, who were involved as research contractors, to leave the country and work in Scandinavia instead. It was very British in the way that the issue was treated as a sort of gladiatoral scientific contest. Whereas in other countries it was a matter for policy action, they didn't view it in such a narrow context. For them the key thing was to take precautionary steps in case the damage got out of hand.

Figure 16 FoE's acid raid 'litmus paper' billboard campaign: this poster was to win the BBC Design Award, 1994

> It meant that Friends of the Earth, which had practically no resources, was taking on people like the Forestry Commission and the CEGB, this giant, multi-billionaire state-within-a-state with all its research capabilities. They were having this scientific argument with us as though it was a head-to-head and we were half of it and they were the other. The fact we were just two people with one typewriter went unremarked.

The *Daily Telegraph* dismissed the acid rain 'problem' as nothing more than 'second-rate science fiction' and compared the issue to what they described as 'the ozone scare' of the 1970s, a prediction by James Lovelock that the accumulation of certain industrial chemicals in the air might deplete the stratospheric ozone layer that shielded life on earth from an overdose of solar radiation. 'We must beware,' the *Telegraph* added, 'of hysterical solutions to complex problems urged by people whose real motive is often hatred of industry and capitalism', a veiled reference to FoE.

To resolve the stalemate, FoE launched their own Tree Survey and with support from WWF produced a kit based on German models that could be used to sample beech and yew trees, two 'indicator' tree species notably sensitive to acidification. Over 3,000 trees were surveyed, most by FoE local group members. Some 69 per cent of beech and 79 per cent of yew trees showed signs of damage. There were howls of protest from professional foresters and ecologists who accused FoE of amateurishness and insisted that these results could not be relied on.

William Waldegrave, then Secretary of State at the Department of the Environment, was unimpressed by this outcry. It was, he said, unfair to criticise FoE for doing what others should already have done. He commissioned the Institute of Terrestrial Ecology to mount a major emergency tree survey. To nobody's great surprise the outcome, published in 1987, closely matched the 1985 FoE survey data.

Controls over power station emissions demanded by FoE and others began to come in during the late 1980s. Though the UK never signed up to the international pact on reducing sulphur emissions, some major power stations were fitted with desulphurisation equipment. Following a bitter and protracted industrial showdown between miners' unions and the Government in the later 1980s, many power stations were converted to low-sulphur coal imported from Eastern Europe, a gain clouded by ulterior political motives. New gas-fired plant was built (and opposed by FoE because of the limited gas reserves and lack of emphasis on energy conservation).

By then the worst of the damage had been done in any case. 'The problem was,' muses Rose, 'that by the time we'd discovered what was going on, it was normal. Only if you go to northwest Scotland or somewhere out of the way can you see completely healthy trees now in the British countryside.'

This point is amply confirmed by 'closed chamber' experiments conducted under all-Europe research programmes, in which air around tree seedlings is filtered to cut out the 'photochemical smog' effect that arises when sunlight reacts with NOx to create a ground layer of toxic ozone gas. These and similar experiments conducted more recently in Denmark, using purified water on enclosed forest trees to mask out acid rain effects, show truly dramatic effects. The trees grow more and bigger leaves which stay on the plant longer and are not covered in fungal blotches, a side-effect of pollution damage. Chris Rose compares the effect of acidification on the landscape to the disappearance of lichens from industrial areas during the Industrial Revolution. 'That transformed the appearance of all the trees, the buildings, everything. But everyone's used to it, it's normal to them now. There's no comparison, so there's no disappointment to get an angle on.'

Charles Secrett, meanwhile, had begun to turn a spotlight onto global aspects of forest loss, laying the ground for a five-year multi-country campaign on rainforests, supported by gifted allies like ERR researcher David Baldock and *New Scientist* environment stringer Catherine Caufield. Most pundits now acknowledge that there was no bigger wildlife or environmental issue in the late 1970s and early 1980s than the rapid and wholesale devastation of tropical rainforests, a habitat that more than any other lives up to the epithet 'the cradle of life'. In terms of sheer tonnage, rainforest contains almost a third of the planet's entire terrestrial biomass. They are also precious reserves of biodiversity – unique genetic data tied up in species and ecosystems – and the origin of many of the crops on which human survival depends.

Figures published by the UN Food and Agriculture Organization during the 1980s drew attention to loss of global forest cover, without replacement, at a rate of almost 1 per cent a year: loss of tropical forest cover was even more drastic than this average figure suggested, amounting to more than 6 million hectares a year in tropical South America alone, with comparable trends showing in central Africa and south-east Asia. Simple loss of rainforest area was not the only issue. The accompanying loss of biomass and species richness from the Equator's fragmentary girdle of green, though impossible to measure exactly, gave still greater cause for concern.

Various reasons were offered to account for this alarming vanishing act: forest clearance to support intensive agriculture or make way for highways, hydro-power schemes and settlements; shifting cultivation and the burning of fuelwood by local people hungry for land and energy; the spread of destructive logging and mining. Whatever the cause, the effects were disturbing to contemplate. Rainforests were not only vital global 'gene banks' and nature reserves but also, crucially, served to regulate world climate and atmosphere.

As understanding of these hidden costs grew in the 1970s, concerned scientists and internationalists like Norman Myers, Jack Westoby and P. W. Richards gave voice to growing disquiet. Their warnings began to make it clear that tropical forests were now a major environmental arena where issues crucially forming the planet's destiny had come to a head.

Secrett and his allies took the view that the organisation's claim to be Friends of the Earth rang false unless it took active steps to prove itself a genuine friend of the rainforest. This was no simple proposition. Secrett explains,

> The goal of the campaign was quite new. Tropical rainforests were thought of then as a forestry researcher's interest. They weren't seen as a campaigning issue at all. Okay, there were concerned scientists and there were the very beginnings of environmental movements in tropical countries around forest issues. But for all sorts of political, economic, social and cultural reasons, they weren't going to catch fire without corresponding pressure in the North.
>
> Nobody believed rainforests could become a major campaign theme for Friends of the Earth. We had to answer the question: How on earth can we campaign on an issue like this in this country when the nearest rainforest is thousands of miles away? Most people said: Forget it, we can't do it. I said, of course we can. We have to just analyse the problem, see what the forces of destruction are, and work out the connections. We needed to build up a coherent overview of *why* these forests were being destroyed and analyse the political and economic forces that funded the deforestation. We could then identify those responsible – the Government Departments, multilateral development banks, logging and mining companies – and campaign to change them. Above all, we had to show, rather as in the Whale campaign,

> why rainforests were so special and important, even if you had never been near one in your life.

While still running the Countryside campaign, Secrett spent a year trying to disentangle these strands, mostly in the evenings and at weekends. When the campaign was launched in April 1985, he had formed a five-year strategy to turn tropical forests into the environmental issue of the decade.

> We set out wanting to build an international dimension into what was still understood as a British wildlife and habitats campaign. Because you don't exist in isolation and whatever you do to protect the habitats of a particular country you've got to also realise what's happening at a global level and the interconnections between countries and their economies and political systems, the drives and needs that lead to degradation and catastrophe elsewhere.

By the 1990s, this approach to bringing global issues home would have become a truism. The mahogany toilet seat came to epitomise the way the tastes and consumer power of the industrialised North wreaks havoc on natural habitats, and people's lives, in the South. But at the time Secrett's proposition was a leap in the dark.

One reason why he was given such unaccustomed leeway and lead time to concentrate on forward planning for this radical shift in pro-environment gut thinking, was that the organisation's helm was taken over in late 1984 by a new Director whose commitment to internationalism and lateral thinking coincided in many respects with Secrett's, Jonathon Porritt. Porritt's communication skills would come to be seen as his main gift to the organisation but his strategic bent was equally crucial.

A schoolteacher by profession, Porritt had been a member of an FoE local group in London from 1974, albeit (by his own account) 'an extremely inactive member, doing nothing much but pay my dues or occasionally turn out for the odd meeting. I liked the idea that it was based on local groups, as this seems to me an important organisational principle behind the green movement in general.'

In the late 1970s, however, his activist commitment had broadened out when he joined what was then known as the UK Ecology Party, later renamed the Green Party. Inspired by novel political developments on the European mainland, especially in Germany, the Ecology Party had been set up in the UK in 1973 by people interested in a brand of global politics that hinged on a systemic change in human behaviour as the answer to the world's ills. It opposed economic growth and the consumer society outright and saw social change as the key to global environmental security.

In Britain, Green politics as a distinctive voting option had little going for it in the 1970s. The scope for advancement enjoyed by small parties in countries such as Germany, The Netherlands and Belgium, which had electoral systems based on proportional representation, was a far cry from the British situation. This and other crucial differences meant that the Green Party's historic role in the UK was to trigger rivalry between the major parties to be the most green, while never becoming fully credible as a separate party worth voting for. FoE had generally ignored the Green Party, which it saw as politically naive and largely irrelevant.

In 1983, in the midst of a world debt crisis that gave way to clear symptoms of global recession, the German Greens, Die Grünen, won 5.6 per cent of the national vote and a place in the Bundestag. This turn of events, combined with fast-growing support for independent environmental groups at home, attracted the attention of strategists in Britain's Labour Party and the left-of-centre Liberal and Social Democratic Parties, who saw it as evidence of growing demand for alternatives to hardline monetarist political doctrines embodied by Margaret Thatcher here and Ronald Reagan in the USA.

Later Thatcher, encouraged by Britain's Ambassador to the UN, Sir Crispin Tickell, would embark on a famous U-turn. From implacably opposing environmental protection as a threat to the wealth of nations, she would turn to a bizarre new role as a roving doomwatch orator. In 1989 the UK Green Party would poll 15 per cent of the vote in European elections. Both these turnabouts would come as a result of an extraordinary surge of popular environmental concern in Britain in the later 1980s.

But as Porritt points out in *The Coming of the Greens* (published in 1988 and co-written with David Winner), 'the Green Party itself was actually slower to respond to this trend than the mainstream parties. Nothing annoyed [it] more than reference to the Party of the Environment.' What the Green Party was slow to appreciate, in Porritt's view, was that for most people the word green defined ideas that were almost exclusively to do with environment. The Green Party, on the other hand, viewed environment as only one aspect of a far more inclusive raft of concerns, encompassing health, education, the economy, international relations, spirituality, lifestyle and much else besides.

Drawing the same distinction between the environment movement and the Green movement in his first book *Seeing Green*, published shortly after his appointment as Director of FoE, Porritt had issued a challenge 'to an environment movement which prides itself on its political neutrality'. He did not, he said, believe such an approach was any longer viable, adding that 'concern for the environment provides as good a starting point as any for green politics'.

Though Porritt would draw fire, as much for politicising the issues to a standstill as for overselling a philosophical reading of them, his ability to interpret environmental topics in a structured yet eloquent way and set them in their true political context had impressed Des Wilson. In 1983 Wilson had inaugurated a series of FoE Green Rallies around the country on key environmental topics. Porritt had represented the Ecology Party at some of these events and had stood out as something of a star speaker.

The rallies had been a big success. Wilson persuaded Ralph Nader, famous for his consumer crusades in America, to speak at the inaugural event in Westminster City Hall then at a string of regional meetings, fronting a panel of speakers that on different occasions included Petra Kelly of Die Grunen and TV botanist David Bellamy. Another familiar figure in the debates was Graham Searle, who had quit ERR and gone to ground in rural East Anglia as a freelance consultant, only to find the Sizewell B controversy raging in his backyard. Now he chaired a highly vocal 'Stop Sizewell B' campaign group.

Porritt's schoolteaching experience may have contributed much to his wake-up-at-the-back-of-the-class presence and his flair for off-the-cuff pronouncements on a wide range of issues. At the root of his personal outlook lay a consistent belief closely in tune with the ethos of Green politics, namely that the quest for environmental justice was intimately linked to a wider social, civil and moral debate. Without social welfare and democratic renewal in the script, green advocacy risked becoming irrelevant to the course of history.

This was not a new idea, nor was it foreign to the founding principles of FoE in Britain. A prominent item in *The Environmental Handbook*, in its 1971 UK version, had been a reprint of a 1967 *Spectator* article by Edward J. Mishan, which declared that 'an extension of choice in respect of environment is the one really significant contribution to social welfare that is immediately feasible.'

Des Wilson began looking for someone to take over as Director in 1983. Porritt's stock was high. He was already a darling of the media, ever-ready with quotable quotes that were delivered with well-spoken charm and evident good sense. In 1983 Steve Billcliffe had announced that he wanted to re-enter party politics and stand as a Labour candidate for Bury St Edmunds at that year's elections. The Board and staff interpreted this declaration as a bar to Billcliffe's continuing in charge of an avowedly non-political oufit. Others regarded these objections as a Trojan Horse, Billcliffe having become increasingly marginalised at FoE.

Jonathon Porritt stood out as an obvious successor, although he had some doubts about quitting his teaching career. In the end he decided the challenge was worthwhile and took the job, little suspecting he was about to preside over a boomtime.

12

Catching the wave

> However confident the leading organisations may be about their prospects over the next few years, the harsh fact remains that it's really very difficult to notch up any substantial progress across a range of environmental issues. In certain cases it may even be true that mere activity is being confused with achievement.
>
> – Jonathon Porritt (with David Winner), *The Coming of the Greens,* 1988

Between 1985 and 1993, membership of the Council for the Protection of Rural England (CPRE) grew from 30,000 to 46,000. Memberships of the National Trust and its Scottish equivalent doubled in numbers during the 1980s, likewise those of the Ramblers' Association and the Royal Society for the Protection of Birds. By 1990, the RSPB's income had climbed to over £22 million and its members numbered an astounding 770,000.

The convulsive expansion of these national conservation groups, all of which operated (broadly speaking) in a long-established pro-nature tradition, coincided with the rise to fame of newer groups like WWF, Greenpeace and Friends of the Earth. A more professional approach to raising funds and recruiting members partly accounted for this surge. But these marketing efforts benefited in large measure from the topical and populist campaigning methods the newer global groups had brought to bear on issues affecting the environment. These newcomers grew too: Greenpeace from 30,000 members in 1981 to 408,000 in 1993, WWF from 60,000 to 227,000 and FoE from 19,000 to 230,000 members in Britain alone. But the larger groups were by no means the only beneficiaries.

Another key feature of the late 1980s was the swift growth of localised and single-issue environmentalism in Britain. Some new groups pursued tricky ethical causes like animal welfare on a national 'media war' footing and some used direct action tactics that verged on defiance of legal means of protest. But most of the new groups were small-scale, low-profile outfits that gently extended selected aspects of the green agenda. Typical among their goals were trials of alternative organic farming or renewable energy production methods, recycling schemes, energy conservation 'cottage industries' and schemes for regenerating derelict urban areas as city nature parks.

Many took advantage of State-backed job-creation grants under Youth Opportunity Programmes or Manpower Services Commission schemes, aimed at reducing unemployment. The UK jobless total exceeded 3 million after the recessionary trough of the early 1980s, combined with drastic reductions in spending on public and local government services and jobs.

Some FoE local groups benefited cannily by job creation schemes. At one point, FoE Birmingham employed over 100 people full-time on various work programmes around the city and its suburbs, a payroll many times larger than that of the national organisation. What these trends seemed to indicate was that the pursuit of practical environmental goals was now more respectable than ever before.

But the gains, though undeniably impressive and valuable in myriad instances, were somewhat of a letdown in the mass. The quality of the environment continued to decline overall despite all that was being done here and there at local level. Nationally, no matter how successfully nature conservation organisations raised awareness and recruited new members, loss of key landscapes and habitat, even of SSSIs, continued apace.

New subsidy mechanisms introduced under the EC Common Agricultural Policy were speeding this process, rewarding farmers for cultivating marginal land previously left to nature. Air and water pollution problems still showed few obvious sign of easing despite a host of new EC directives on air and water quality standards. A new menace was emerging in the proliferation of grandiose highway construction projects, and unprecedented rises in car ownership and road traffic.

The global situation was similar. UN estimates suggest that at least 2,000 new national environmental groups were founded throughout developing regions during the 1980s. In response to hard-hitting environmental campaigns waged in the North and South, major development banks, bilateral aid agencies and intergovernmental bodies went to increasing pains to declare the best of environmental intentions and place sustainable development at the head of the global investment agenda. Yet it remained obvious that consumer, trade and political arrangements originating in the geopolitical North, were still driving the destruction of key environments in the South, most obviously the tropical forests of Latin America and Asia. 'The main reason why tropical forests were being cleared at such fantastic rates,' says Charles Secrett, 'was because of the way Northern companies operated, the way Northern consumers expected their demand for forest products to be satisfied, and the way their governments channelled aid to tropical zones.'

These forces of destruction worked in direct and indirect ways. The North's wealth was driving the destruction of the forest through trade without responsible stewardship. At the same time the conditions of extreme poverty in the South that led to further forest loss were perpetuated by wealthy ruling elites who monopolised productive land and its resources. That meant vast numbers of people in Central and South America, central Africa and south-east Asia were deprived of land and livelihood.

All too often, the only recourse open to the poor was to invade protected forest lands illegally and clear them for agriculture, simply to survive. Or in some cases, notably in parts of Indonesia, the urban masses were being encouraged to settle forest lands as a way of colonising disputed border territories. In Brazil, resettlement was used as a safety valve to relieve hopeless overcrowding in cities or dustbowl conditions in rural drylands. Secrett explains,

> What we tried to do over five years, working with others in this country and overseas, in the North and the South, was to build a tropical forest campaigning movement that clarified the causes of forest destruction but also helped people in this country and the developed world see their role and understand their responsibility to help protect these irreplaceable ecosystems.
>
> The FoE Tropical Forests Campaign took as its starting-point the idea that in order to protect tropical forests we had to change the way we did things in this country, in Europe or in Japan or North America. The trading and political blocs of the North were the motors of the devastation. We needed to reform the way we traded, the way we dealt with development schemes, the role of institutions like the World Bank and the high street banks. And we had to reform our own consumer expectations.

The first ingredient in the campaign, launched in 1985, was to establish these connections in Northern consciousness. Says Secrett,

> I felt the best way to do that was to open people's eyes to all the benefits we obtain from tropical forests, things we take totally for granted yet without the tropical forests we wouldn't have. Products like medicines, food, timber and other industrial goods. In a way it was about establishing tropical rainforests as a symbol for all the environmental movement stood for. We wanted to make Protect the Rainforest as much a rallying cry for the eighties and beyond as Save the Whale had been in the seventies.

One of the first publicity stories that helped get the campaign off to a flying start in Britain drew on a very unlikely sounding connection with a cultural totem, the Great British Breakfast. It showed how virtually everything in the classic British breakfast originated from tropical forests, beginning with packaged cereals like Coco-Pops, cornflakes or Rice Krispies. Their main ingredients – chocolate, maize, and rice – all began as tropical forest plants taken into cultivation.

Fruit juices, marmalade, sugar, coffee and tea were equally traceable to tropical forest origins. Even bacon and eggs owed some debt to interbreeding with tropical forest species as farmyard species of pig and chicken were 'improved' to satisfy mass markets. This way of encapsulating everyday connections between tropical forests and British consumers struck an arresting chord. The publicity, and public response, were tremendous. Then medicines were given the same treatment. Secrett elaborates:

> Go into your high street chemist and you've got a one in four chance of buying a medicinal or cosmetic product based on tropical rainforest plants. Then look at the economy, the industrial side, look at rubber, the classic example of a rainforest species that has been an absolutely critical industrial product behind the creation of millions of jobs worldwide over the last fifty years. We began to open up perspectives on this and other valuable goods, the dyes, the chemicals, the waxes, the gums, the fibres and the timbers and minerals that originate in tropical forests. Without the forests or the lands they grow on being managed on a sustainable basis, these goods would not exist. We wanted to show that it wasn't just a matter of looking after the environment, it was also about safeguarding a sustainable way of life and the continued supply of items we need or want.

The main launch of the campaign concentrated on the most direct and obvious linkage between consumers and rainforests, the timber trade. Research showed that Britain was one of the biggest consumers of tropical timbers in the world. Other key marketplaces were mainland Europe, North America and Japan.

> The idea was to start with the timber trade in this country, pointing out opportunities politicians and consumers had to help shape markets, restructure the trade and create a stable basis for sustainably managing tropical forests. Then we looked to export that campaign to Europe, to America and to Japan over three years. So we began by analysing the British Tropical Timber Association, the main logging and processing companies, the ports where timber flows were coming in, the distribution networks, and we tracked it right back to the retail level and so to people's everyday lives.

The result was the Tropical Forest Products List, which ran to almost forty pages, prepared by Secrett and a committed volunteer, Simon Counsell. Following the model developed for FoE's Whale and Endangered Species campaigns, the list named UK retailers selling rainforest products from non-sustainable sources, explaining what the woods were, who was selling them and where. A code of conduct for the trade was devised and the Timber Trade Association was vigorously lobbied to accept it, to help conserve the forests, maintain a reliable timber flow and show that a decent economic return for tropical nations did not necessitate killing the golden goose, the forests themselves. These materials became a basis for local group campaigns aimed at motivating consumers to use their spending power to affect the retail marketplace.

Chris Church, local groups coordinator at City Road at the time, describes how these activities developed, first through local surveys of furniture and DIY stores which were fed back to the centre for incorporation into the Product List. Then some of the groups moved on to picket selected shops and flypost products such as furniture, doors or toilet seats with small hazard warning stickers bearing slogans such as 'CAUTION! THIS PRODUCT KILLS PARROTS' or 'A MONKEY LOST ITS NUTS FOR THIS PRODUCT'.

'The trade's initial reaction had been to ignore Friends of the Earth,' Church recounts, 'but as the campaign developed, their attitude became increasingly schizophrenic.' The more aware traders saw that unless something was done they would ultimately be deprived of their own stock in trade. They preferred a voluntary (as opposed to statutory) approach to management agreements and a system of labelling of the kind that green consumer movements would adopt a few years later.

'Generally the industry was surprised by the success of the campaign,' says Church. 'In 1986 some of the less enlightened companies got increasingly aggressive towards us and became totally uncooperative.' Much of the debate within the trade was conducted in the columns of the *Timber Trades Journal*, which gave the campaign extensive coverage.

The debate heated up noticeably after the next round of campaign materials came out, the aforementioned stickers and a mini-poster featuring a mahogany toilet seat, headlined 'Every year 11 million acres of rainforest are destroyed for the sake of convenience.' The poster was distributed widely, used as an advertisement and sent to every MP. It urged everyone to let shopkeepers know about their disapproval of trade in such products. Notified by FoE Camden that they intended to picket the chain's flagship London branch, Habitat contacted the central office and asked for advice on what they should be doing. Habitat saw which way the wind was blowing and removed offending products from its stores. The campaign was beginning to make a difference.

At the same time as spurring local action, FoE was lobbying the British Government to seek legal changes and controls over the range of species the timber trade could retail, as well as statutory incentives for traders to move their business onto a more sustainable basis. A model European Directive was drawn up to achieve the same goal across Europe, which quickly attracted the support of many MEPs. In harness with other international environmental groups, the organisation was lobbying the newly founded International Tropical Timber Organisation with the same goal in mind.

The next phase of the campaign meant undertaking a European, then a North American, then a Japanese analysis. As the strategy unfolded over time, the campaign drew attention to other destructive economic pressures besides the timber trade: mining, private banking, transport and agriculture. In each case, meticulous research, packaged in hard-hitting campaign reports, uncovered the Government policies and companies responsible for the destruction of the forests, identified clear campaign targets, recommended solutions, and then mobilised consumers, shareholders, voters and politicians to press for the necessary changes.

> The third strand was to look at another revenue stream, channelled through the international development agencies. Taxpayers' money was being used as development aid to fund destructive projects. Dams and highways are the classic examples. Here we were able to strike up an alliance early on with the Environmental Defense Fund in America, who were running a very successful campaign against the World Bank's role as a major funder of dodgy rainforest development projects. Mostly these were transmigration schemes, inappropriate plantation forestry programmes, destructive logging drives or dam and highway projects.

With EDF support, the Sierra Club had published a bombshell indictment of the World Bank, *Bankrolling Disaster*, that zeroed in on destructive hydro-power projects in Amazonia funded by the Bank. Other texts that influenced the US debate included *The Hamburger Connection*, which drew connections between forest clearing for beef ranching in Latin America and the activities of giant fast-food burger chains.

FoE concluded early on that the evidence for this connection was slight enough to let the burger chains off the hook, but the criticism levelled at the World Bank by the EDF with the support of FoE and other key groups found its mark. In 1987, Bank President F. Barber Conable announced a complete restructuring of the World Bank to ensure that it would not back environmentally suspect projects in the future. The stakes were evidently rising.

Working with Survival International, WWF and Oxfam, FoE began to take the campaign to the South, establishing supportive links with emerging national activist groups concerned about rainforests in Sarawak, Brazil and Malaysia. The British and EC development aid programmes were scrutinised along with streams of development funding channelled through Britain's Overseas Development Administration and other bilateral agencies into the World Bank and the regional development banks. Secrett shuttled between all points of the compass, meeting with representatives of environmental groups in the USA, throughout Europe, in Japan, Malaysia and Brazil, explaining FoE's campaign and arguing that, despite the scale of the problem, tropical deforestation was an issue that the movement could work on effectively. New groups and alliances sprang up around the world under a Protect the Rainforest flag. Says Secrett,

> That was the five-year goal, for all these strands to come together in a global campaign, so that at the end we could all picture the intimate connections between individuals, communities and nations all over the world and the forces that led to the devastation of the rainforest – and be confident that there was much that citizens working together could do. The huge political and economic inequalities between North and South lay at the root of so much of what we had to change. That was the plan, and by and large it worked.
>
> Three years after the launch I could go out and address public meetings almost anywhere in the UK and talk to people who had come to hear about the global debt crisis and tropical deforestation on the same ticket. Who

Figure 17 Chief Paiakan speaking to Xuante and Kayapo people at the Altamira Conference, Brazil, 1988 © Susan Cunningham

> would have believed that possible when we launched? That, as much as anything, was the indicator of how successful the campaign was. And we exported this campaign model everywhere, all that has been done since on tropical forest campaigning is based on one or other aspect of that campaign strategy.

Indeed, in 1987, when Secrett became campaigns coordinator and passed the tropical rainforest campaign on to Koy Thomson, several more successes were quickly chalked up. A high-profile international attack on Scott Paper dissuaded that company from pulping timber from Indonesian rainforests inhabited by indigenous tribes. This campaign built on an earlier success, carried forward by Secrett and Counsell, to discourage the Coca Cola Corporation from converting forests in Belize into citrus plantations. By the time of Thomson's departure in 1990, the tropical rainforest issue was firmly established as a mainstream environmental concern.

When Tony Juniper joined FoE in 1990, the international aid institutions, transnational corporations, tropical timber traders and commercial banks, though firmly on the defensive, had begun to fight back. They suddenly showed a concern for tropical forests that was – to the uninitiated – almost credible. Timber traders queued up to claim that their wood came from sustainable sources; aid agencies emphasised respect for indigenous landrights and sustainable development; transnational companies developed codes of conduct and claimed that their real interests lay with the poor and the environment. The counterattack aimed at preserving the status quo had obviously started, and it had to be dealt with.

In 1991, FoE published *Plunder in Ghana's Rainforests for Illegal Profit.* Researched and authored by Tim Rice, this landmark report exposed a web of corruption and malpractice that underlined the destruction wrought by a World Bank economic 'recovery' programme, and revealed how the newly greened tropical timber trade had engaged in fraud, smuggling and other malpractice to rob one of Africa's poorest countries of millions of dollars' worth of foreign exchange. This salvo was followed by further investigations into aid projects, timber firms and oil companies working in remote forest areas. More successes followed. The African Development Bank withdrew from a disastrous coffee growing project in the Central African Republic; British timber traders were linked with illegally exported timber from tropical countries; and a boycott aimed at British High Street banks took the campaign once more to towns throughout the country. After several months of campaigning pressure, British Gas withdrew from an oil development project in the Amazonian forests of eastern Ecuador, and a joint campaign with WWF resulted in the World Bank undertaking not to fund any further logging in primary tropical forests.

By 1993, enough material had been amassed to launch one of the hardest hitting campaigns of all: Mahogany is Murder. Based on initial research carried out by George Monbiot, it implicated the British timber and furniture industries in the murder of Brazilian Indians because of their use of mahogany taken illegally, and very destructively, from Indian reserves in the rainforests of the Amazon basin. More investigations, direct action, a consumer boycott and an advertising campaign piled the pressure on the mahogany traders. Imports to the UK plummeted. In 1994, lobbying by campaigners at the ninth meeting of the CITES Convention came very close to getting the international mahogany trade regulated. The struggle continues.

Tony Juniper sees the arguments as largely won but the battle far from over:

> We have been up against some pretty fundamental forces – the transnationals' objective to expand and liberalise trade, the aid agencies' promotion of the donor governments' agenda to increase exports to developing countries, and the liquidation of resources to service debts have enormous momentum behind them. But these issues have now taken on lives of their own and campaigners all over the world are now able to focus on the underlying causes not only of environmental destruction, but also of social injustice and economic domination in ways that would not have been possible without the high profile of the tropical rainforest issue. This campaign will be seen in future as marking a critically important phase in the struggle for socially just and sustainable development.

Secrett concludes that the campaign's most crucial payload was

> the way the campaign evolved throughout the Friends of the Earth International network. Our approach was exactly what the embryonic environmental movement in the South had been hoping to see in the North. As they well knew, you can't address fundamental environmental issues like these unless you have an overtly political, social and economic dimension to your campaigns, because that's where the root causes of why we're messing up the planet lie. We were in this together, North and South complementing one another.

On a more mundane note, appeals that went out in Britain in connection with the Protect the Rainforest campaign had been a fund-raising breakthrough, netting more donations than any other campaign since the Save the Whale bandwagon of the 1970s. The local campaigns had evolved from picketing the retailers to persuading local authorities, architecture practices and other large users to phase out the use of hardwood timbers from non-sustainable rainforest sources.

And the timber trade itself was to some extent placated when FoE brought out the first *Good Wood Guide*, also compiled by Secrett and Counsell, listing those companies that supported the campaign in various ways as well as those that opposed it. It was the first instance of even nominal endorsement by the organisation of companies noted for good environmental practice. The Guide was launched in 1987 by local groups.

A showbiz bonus was forthcoming when film director John Boorman, canvassed by Jonathon Porritt, agreed to allow local groups to leaflet filmgoers who turned up to watch his movie *The Emerald Forest*, which was shot largely on location in the Amazonian rainforest. Tens of thousands of leaflets were handed out, outlining key rainforest issues under an appeal message written by Boorman. The shot earned £14,000 and the groups recruited many new local activists and volunteers.

Another organisation that was learning fast about the box-office appeal of rainforest issues was WWF, which launched its own tropical rainforest campaign not long after Secrett's. WWF had recruited Chris Rose to help steer this effort when he quit FoE in 1985. WWF International's chiefs had in fact provided for *two* rainforest campaign launches.

One focused on threatened primates of the rainforest, in the time-honoured WWF tradition of fronting appeals with charismatic animals. The other, which some Fund principals eyed with grave suspicion, led with a much more complex and technical case about the importance of rainforests in the big picture of the Earth's plant life and climate as well as their social and economic significance. The primate campaign was held on standby in case this science and sustainable development focus left the public cold. To the surprise of the doubters, the campaign grossed more donations than any other WWF appeal to date.

An important conclusion could be drawn from this surprise success. It was that the public, thanks not least to WWF's own innovative environmental education activities, had grown a good deal more knowledgeable about scientific and social issues than many veteran wildlife campaigners gave them credit for. The role of the media, particularly TV natural history and science documentaries, in this shift had also been crucial.

One man unsurprised by it all was Jonathon Porritt, who had seen the sea change occurring at first hand, in the classroom. Porritt ran FoE for six years from 1984 to 1990. He now looks back on it as 'a wonderful time, six years with everything changing from day to day'. His reading of the group's achievements in that period fixes characteristically on an overview rather than specifics.

> First of all, Friends of the Earth was able to give voice to some very radical ideas that allowed it to reach practically everybody in the country. There were always one or two politicians or vested interests or businesses who still thought it was too wild and whacky even to contemplate. But by and large we managed to take radical messages wherever we needed to take them. And I think that was a genuinely new departure for an organisation of that kind.
>
> I guess the other key thing Friends of the Earth managed during that time, was not to lose sight of the fact that we had to achieve through the interventions of people. People had to feel a sense of ownership of the issues.

Tom Burke sees the Porritt era in a different light. 'One of the things Jonathon did for Friends was to preserve its presence,' says Burke. 'He caught the wave brilliantly and was extremely good at communicating the issues. But the fact he didn't have a Board which was able to hold him to account meant he didn't always get the other, campaigning bit right.'

Figure 18 FoE Director Jonathon Porritt protests against Third World Debt, 1990
© Rob Hadley/FoE

Burke's is arguably not the most detached of views on this score, considering the historic mantrap he had fallen into over local group representation on the Board. Nor is it borne out by the organisation's record of tangible campaign successes during the late 1980s (not least the rainforest campaign) and the Board's more than adequate part in them.

But it does partly lift a veil on currents of opinion within the national office that made it hard for Porritt to gain initial acceptance for his cerebral and politically conscious response to the issues of the day. Chris Rose, who left the organisation shortly after Porritt's arrival, now says he felt that some FoE campaigns under Jonathon Porritt were 'pretty ineffective' by reason of Porritt's tendency to 'try to run a pressure group as if it was a political party. He wanted us to have lots of policies.'

Porritt had just published his first book, *Seeing Green*, a best-seller in its genre that rounded up his Green Party experiences. It contained some harsh criticisms of the political establishment and a strong plea for ecopolitics. Porritt admits,

> That was embarrassing, because the book had been written by the time I joined the organisation but hadn't come out. I tried to warn Des Wilson and the Board about this. Though the book wasn't party political – at least, it had something bad to say about all the parties – I knew they'd have to weather some political flak because of it.
>
> There was only a very small staff involved permanently in those days, about fifteen all told. Some of them spent quite a long time feeling that I was using my position as Director of Friends of the Earth as a front for Green Party activities.

When Porritt joined FoE, in common with several other environmental groups in the recessionary years of the early 1980s, it was just emerging from a financially depressed and seriously demotivated phase. The US chapter of the organisation had just suffered a traumatic breakup caused by splits between factional groups operating on the West and East Coasts. The Washington cadre had emerged as the dominant party but the costs were high. Even the organisation's founding father, David Brower, had declared a curse on both houses.

Jonathon Porritt takes a sanguine view now of his stormy reception at FoE. He feels that disputes over territory are more or less endemic to campaigning outfits:

> There are two schools of thought. One is that in an ideal world such organisations should somehow cohere, have a focus which would stop those kind of debates going on. I subscribe to the other school, which is that if you're recruiting people who are fantastically committed, living mostly on their nerves or caffeine or some other artificial stimulant, working eighteen hours a day, then nobody should think any management system is going to soften the edges of all those people and create some sort of bland homogeneity in the middle. So you've got to live with that, make that work.
>
> It's a bloody nightmare. I'm not trying to romanticise six years of being Director of Friends of the Earth, I can tell you. But equally I don't like it

> when people write it off as an organisation habitually riven with internal disputes.

Board member Iris Webb feels that a more acute malaise was at work within the organisation in the early 1980s, as FoE campaigners came under growing pressure to prove their individual worth and professionalism:

> For a long time it was very male-dominated and macho. It got to the stage where it deformed people's psyches. A lot of campaigners become egotistical and obsessed to the point where they couldn't sustain marriages or steady relationships. I don't know if they were conscious of it, but there was an association between being aggressive, obsessive, not supportive, with being a good campaigner.

Porritt believes that an infallible antidote to campaigner blues always lay within reach: success. It had the power to direct wayward energies from internal onto external targets where their effect was far more beneficial. 'When things are going well you might have just as much disagreement but everything goes outwards,' he argues. 'You've got this fantastic multiplication of energy going out. If it goes badly you've got a lot of it coming in and if it goes really badly the organisation implodes.'

Some issues never go away. In its continually evolving work on nuclear energy issues, FoE still had an edge on a key contemporary concern of undeniable relevance to the fate of the entire movement. The public at large still saw the organisation above all as a reassuring presence in the increasingly crowded, embattled and confused nuclear arena.

Patrick ('Pad') Green, head of FoE's energy campaign team, takes the view that the main distinguishing feature of the organisation's work in this area has been steady orientation not on problems but on solutions:

> If you look back over twenty-five years, you can see that Friends of the Earth's energy campaigning was always ahead of its time. While we've been opposing particular practices, within a wider policy context we've been putting forward solutions. And not naive solutions, either. People are always criticising opponents of nuclear power as naive. But you know the bloody thing exists. So you've got two choices, a choice about how you deal with the crap they've created and a choice about whether you let them add to it.

Green feels that other groups, particularly in mainland Europe, too often take a stand-off position:

> You know, we won't talk to you about the mess you've created until you stop producing it. That I think is naive. During the ten years I've been involved, I've never been interested in talking only to the converted. At one level there are reasons why you do that, to inform and empower people. But in terms of changing the world I want to talk to someone who doesn't agree with me.

> There's a lot of talk nowadays about solutions-oriented campaigning and I can't help smiling when I hear it presented as some great new thing. But the work I've done and my predecessors – Simon Roberts, Stewart Boyle, Walt Patterson and Czech Conroy, all of them – have done has always been about solutions, trying to get hold of sustainable means of producing energy. Without arrogance, I think we can claim that we've been there, done that and written the book on solutions-oriented campaigning!

Pad Green became involved in nuclear issues campaigning for FoE in October 1985. At university he had studied biology and genetics and had undertaken a third-year project on health and safety in the nuclear industry, at the suggestion of a tutor, Ian Gibson. The Sizewell B Inquiry was ongoing and Gibson mentioned that a leukaemia cluster had been notified at the existing Sizewell plant. Green recounts,

> I spent a year delving around Sizewell, coming down to London, going to the CEGB, having meetings with their senior medical officer and getting the flannel they give you. Everything was going fine till I started wandering around the beach with a geiger counter which I'd borrowed from the university.
>
> On the Sizewell perimeter fence I found there's a radiation hot-spot where you get the gamma shine directly from the reactor building. What the inspectors do to make it okay is average perimeter readings out for public consumption. There was a row between the CEGB and the University. Ian Gibson wrote a piece in the *Times Educational Supplement* about it.

Green went on to do a Master's degree at Aston University, looking specifically at radiation risks. 'I knew,' he says, 'that I was going to do something with it but I still wasn't sure quite what.' Green's Master's thesis examined the basis of radiation safety standards. It looked at the regulatory organisation that set the recommendations, the Nuclear Installation Inspectorate, and revealed that its information was all drawn from the industry itself without independent checks. The work began to attract the interest of environmental groups and Greenpeace used it in evidence to the Australian Royal Commission hearings on bomb tests in the Pacific in 1984. By 1985 Green had started doing consultancy work for both FoE and Greenpeace.

Through a union organiser friend, Tony Webb, he met MP Frank Cook, whose Stockton North constituency in Cleveland was one of the sites then being proposed by NIREX, the nuclear industry's radioactive waste management body, as a suitable location for an underground nuclear waste dump. Webb, Cook and Green had set up a Radiation Victims' Round Table, a House of Commons initative to seek a common campaigning agenda between environmental groups, trades unions, doctors and lawyers. FoE and Greenpeace were also involved. Green adds,

> At that time I was also thinking about a PhD, to give me time to campaign and do the research I needed for campaigns. I was initially going to be funded by Greenpeace and the National Union of Seamen, and Friends of

> the Earth was going to top the funds up. But that all fell through when Greenpeace had some internal problems and the UK Board of Directors was folded. So from October 1985 on I was mainly doing consultancy work for Friends of the Earth.

There was much controversy at the time about the risk of radioactivity and the inadequacy of radiation safety standards. Says Green,

> That debate had polarised, government scientists on the one hand, dissident scientists on the other, and no middle ground. The environmental groups had shied away from this issue because it involved very technical information on risk management, it was very easy to lose your reputation if you got it wrong. I was looking at the people who set the standards, at what their own information showed. I basically argued that those data were perfectly sufficient to prove that the risks were higher than the standards implied.

On the evening of 26 April 1986, Green was in the City Road office waiting to meet Stewart Boyle, the Energy Campaigner of the day. Green recalls,

> It was a Monday about six o'clock. Stewart and I were going to go to a meeting. Chris Church walked in and said he'd just heard that a Russian reactor had blown up, or something of the sort. We all found it intriguing. We got atlases out and looked it up, wondered what it meant. We were more interested in Sellafield, which at that time had been leaking virtually every other week. After about two hours, the phone lines went bananas.

A nuclear reactor at Chernobyl in the Ukraine had exploded, hurling 5 tons of fuel and 50 million curies of radiation into the air. It was the accident they said could never happen.

13

Boomtime

Fire in heaven above:
The image of Possession in Great Measure.
Thus the superior man curbs evil and furthers good,
And thereby obeys the benevolent will of heaven.
– *Ta Chu/Possession in Great Measure*, from *I Ching*

The Chernobyl catastrophe killed thirty-two people outright and left 499 seriously injured (according to Soviet figures). More than 150,000 people were made homeless as the neighbourhood was evacuated. Some 600,000 people were exposed to potentially harmful levels of radiation in the near vicinity of the blast. By 1990 3,000,000 people across a wider area were still under medical supervision and up to 25,000 of them were likely, according to the radiologist's sombre version of the law of averages, to develop radiation-induced cancers over time. No one can really know the number that will eventually die.

In the immediate aftermath of the event, however, world attention focused on a radioactive cloud, the plume from the reactor explosion, which passed over the Ukraine, Byelorussia, Finland, Scandinavia, Poland, Germany, France and Britain, as it carried the danger many hundreds of miles from its source. Animals and food crops over a widening down-wind band of northern Europe were becoming contaminated by the fallout but British Government scientists issued reassuring announcements that the cloud would not pass Britain's way. Even if it did, they said, effects would be insignificant. Pad Green recalls,

> The cloud hit Britain the Monday after, on the Bank Holiday weekend. All the Ministry of Agriculture officials were off playing golf, nobody was there to deal with it. Pandemonium broke out. Our phone lines were jammed solid with farmers, vets, all sorts of people desperate for some sort of advice. We were just giving out commonsense guidance. I remember talking at least two people out of emigrating.

Iris Webb also remembers the excitement of that long weekend. 'When Chernobyl blew, FoE was the main source of information about what to do.

I was the junior on duty on the Tuesday, the phones were hot all day. I have to say it was wonderful for us, we had an immense feeling of being involved in history.' Green adds,

> Friends of the Earth got incredible coverage in the month after Chernobyl, we really cleaned up. It was just the right time to go on the nuclear stuff. In terms of information on the risks of radiation it had become a certainty that these were greater than anybody said. It now seemed just a question of *when* they were going to make the necessary changes and how we were going to prompt that.

Green and Boyle were called in repeatedly by TV news programmes to give informed assessments of the situation. The organisation grew in public stature with each passing day and Green became a 'sort of semi-official' member of the City Road staff. He made a deal with Jonathon Porritt to receive modest consultant funding for a year to carry on – which also enabled him to start his PhD. A major donation from Godfrey Bradman enabled FoE to buy equipment and set up a mobile monitoring unit to keep track of the Chernobyl fallout. Says Green,

> We went to North Wales, Scotland, Cumbria and other areas, to see how much of the stuff had got into farms, into milk and so on. It showed the public that environmental groups were beginning to produce their own data to challenge that of Government and the nuclear industry. That was a critical part of the evolutionary shift. Friends of the Earth had always tried to do its science homework but now the public saw the proof. In the perception of Government, industry and the regulators, that made a difference, they couldn't dismiss us.

Within a year of Chernobyl, the National Radiation Protection Board had admitted that radiation risks were about three times higher than provided for by current safety standards, bearing out Green's argument in spades. The risk management and cost implications of this shift were profound in terms of additional safety provisions and technical improvements now due throughout the industry.

By the time Pad Green's temporary contract became a full-time consultancy, in 1987, a spate of other new appointments was underway, along with a series of new responsibilities, titles and administrative support structures, as revenues flowed into the coffers in amounts previously undreamt-of.

Credit for the organisation's response to the Chernobyl aftermath was part of the reason for this bonanza. Fundraising know-how had also kept pace with growing public support for the campaigns. Press or direct-mail appeals now brought in more than the running costs with enough on top to support a major recruitment and expansion drive. An immediate outcome of this growth was that the City Road premises soon began to burst at the seams. The move to the organisation's present headquarters, a converted warehouse in Underwood Street, a stone's throw from City Road, was hurried through in 1987.

Of all the new recruits to FoE during its dash for growth, none would turn out to be more crucial to the organisation's immediate destiny than Andrew Lees, the young botanist whose work as a local activist in Norfolk, Wales and elsewhere had already marked him out as a rising star. A fellow campaigner in East Anglia, Iris Webb, was one of those who put his name forward to head the Countryside and Agriculture campaign. She says,

> I'd seen him in action. He was a wonderful person who loved the Norfolk Broads. I saw him as a bit of a shark, a lovable shark. At his interview he just blasted everyone else out of the water, knew the issues far better. Though he passed a few typically bolshie remarks that made some of the interview panel wary of him. In the end he got a six-month probationary period in the job to see how he fitted in.

Webb found Lees waiting for a verdict in the basement at City Road.

> He was pale and shaking. He said, you've got to tell me. I said: you've got it, *but* there are lots of buts. And he hugged me for at least a minute. I realised he was crying. He said: if I hadn't got it, I don't know what I would have done. Till then, I didn't realise how badly he wanted to work for Friends of the Earth. It was literally as if his life depended on it.

This private drama would come back to haunt Webb's memory in later years. The driving – and driven – nature of his commitment to the cause would make Lees one of the most charismatic advocates for the environment in the history of the movement. But it would also lead, during the early 1990s, to spasms of office in-fighting and territorial friction. And the Andrew Lees drama would ultimately draw to a tragic finale.

For now, however, he revelled in the new scope his appointment offered to make waves. One thing his new colleagues found was that Lees, for all his flamboyance, was a stickler for sound fact-finding and credible science. Another Lees keynote was that whatever issue happened to come his way, it always somehow took on an aquatic dimension. When Blake Lee-Harwood first went to work for Andrew Lees, at the end of June 1987, the campaign Lees was running was still called the Countryside and Agriculture Campaign, although he had, says Lee-Harwood, 'almost completely diverted it into a Water Campaign. He hadn't been given a straight mandate to call it that but he liked to refer cryptically to rivers as "the moving countryside" so we were effectively running a Water Campaign. About a year later, the organisation allowed the Water Campaign label to stick.'

Lee-Harwood had completed a course in aerial pesticide spraying about a year before he was introduced to Lees by a mutual friend. Lees offered him the chance to be a full-time volunteer. He would help develop campaigns on sprays and on following up EC directives on permissible levels of pesticide residues in river and drinking water. 'At the time I was selling second-hand cars,' says Lee-Harwood. 'I said to myself, yes, this is more fun than that, I'll do it.'

Chemicals on which the campaign mainly focused were dieldrin, aldrin and related insecticides in the organochlorine family, nicknamed the 'drins' by water campaigners. They left residues in the environment long after their application, and they had the potential to accumulate to lethal levels in animal tissues.

Evidence suggested that the 'drins' were a potential cause of cancers in mammals and of high incidental kills of wildlife beyond their intended range of insect pests. This had led to new EC restrictions on the levels of pesticide residues permissible in drinking water. Privatisation of Britain's water utilities was on the cards at the time and Lees wanted to gear up against any official move to relax the rules on water quality aimed at making the sell-off a tastier investment prospect.

The tactical thinking behind the initiative typified the Big Idea angle that Lees gave to most of his campaigns, as another member of the Water Campaign team, biochemist Mary Taylor, recalls.

> On drinking water and water privatisation, the Big Idea was that the quality of tapwater reflected what was going on in the environment. Also that it was a big window of opportunity for a lot of press interest. Start getting people worried about tapwater, then that would lead back through the pipes to the sources which needed better protection. To cut pesticide use, for instance, maybe we could look at numbers of breaches of EC pesticide standards, expose Government motivations and try to enforce or tighten standards.

In this sense, at least, the campaign was still geared to agriculture but the drinking water angle was a brilliant piece of lateral thinking, as events would later prove.

But before the work got fully underway, a yet more unexpected turn of events catapulted FoE onto the world stage and Lees into the limelight as a master-tactician. It all began in early 1988 with a telephone call that Blake Lee-Harwood received from an acquaintance who edited a small-circulation but highly influential inside-track magazine for Africa. This contact had excellent political contacts in many parts of Africa, relates Lee-Harwood.

> Somebody in the Nigerian Government had told him they'd discovered that somebody had just dumped a load of toxic waste in a tiny fishing village called Koko. And because I'd helped him with some research in the past, he had said to his Nigerian contact: I know some guys who work on the environment and chemicals, I'll ring them, maybe they can help.

The contact that was made between FoE and the Nigerian Government led to actions that uncovered an international waste-dumping scandal and triggered a detective hunt and ocean-going chase that the world's media followed breathlessly for weeks and months. At the centre of all this attention was a ship, the *Karin B*, that became a tabloid headline writer's dreamboat, the infamous Ship of Doom. Lee-Harwood takes up the tale from the day he told Lees about the Koko dumping rumour.

> From then a series of very rapid direct exchanges happened between ourselves and the Nigerians. Then a swift decision was made that we would fly out to Nigeria. We'd be met by certain key officials who would then take us to Koko to investigate. We rounded up and took out two consultants, a radiation expert and a toxicologist. There were going to be those two, plus Andrew and Charles Secrett. Andrew also had the inspiration to ring up Independent Television News and ask them if they'd like to supply a reporter and camera crew to accompany them to the site. He sold it to them and ITN agreed.
>
> They travelled at high speed over from Lagos in a fleet of official cars, accompanied by some fairly heavy *hombres* wearing dark glasses. Needless to say they arrived without getting much hassle at military checkpoints. Andrew did his stuff with the TV crew, nosing around and asking around about when and how it had all happened. Charles took photos and the two experts took dip samples and did a radiation sweep.

What they found was 3,800 tonnes of chemical waste, none of it radioactive, as tests back in England subsequently proved, but all of it highly toxic. It was stored in rusty oildrums stacked three or four deep in the backyard of a farmer called Sunday Nana. The only protection was a rickety fence full of holes. Lee-Harwood says,

> Whoever it was had just rolled it off a boat down gangplanks in this sleepy tropical port and stored it in the yard. Nana told Andrew that the Europeans who dumped the stuff gave him the equivalent of £76 to keep it for them for a year. Of course there was never any intention to retrieve it. When they were there, Andrew said you could hear the barrels of solvent wastes groaning as they expanded in the sun. Left there long enough, they would almost certainly have exploded and caused a major disaster. And local people had no idea what was in them. The villagers were emptying drums onto the ground to take home to store water in. School kids were using the place as an adventure playground.

Most of the barrels appeared to have come from Italy, or at least from a transit point in Italy. Some had 'A GIFT FROM THE ITALIAN GOVERNMENT' emblazoned on their side. The Italian touch would later fuel fervid speculation about a Mafia connection.

The campaigners returned and broke the story. There was a press conference and ITN put the story high up on their main evening news. 'Suddenly we had an International Toxic Wastes campaign on our hands,' says Lee-Harwood. 'We had to become experts on the subject in a hurry.'

The Italian Government protested that it had nothing to do with the incident and claimed to be as much a victim as anyone else. The public saved its concern for the people whose lives had been threatened by this horrifying example of criminal and racist irresponsibility on the part of big industry. The Koko tip-off had been a lucky break. How many more such abuses were going by the board in other developing countries? Italy and seventeen other Mediterranean states had ratified the 1976 Barcelona Convention and signed up to strict limits on the dumping of hazardous

Figure 19 Andrew Lees, Charles Secrett and colleagues examine toxic waste dump site in Koko, Nigeria © C. Secrett/FoE

wastes at sea or by its margins. Was this how the polluters were interpreting their duty of care?

'We stayed in contact with the Nigerians for some time,' recalls Lee-Harwood. 'There was a lot of poking around, false leads and so on. But ultimately more of the story did unfold, and not just because of our work. Lots of other people were working on the story by now. In August 1988, we learned that the waste had been shipped out of Koko on the *Karin B*.'

Sensing a story, Andrew Lees signed up for Lloyds of London's Tracking Scheme, which (for a price) enabled subscribers to locate registered vessels anywhere in the world by satellite and radio. Soon a faxed report arrived that the *Karin B* had reached the coast of Africa. Then it was tracked returning north as far as Cadiz. The ship lay at anchor off Cadiz for days. The question was, was it going to turn right into the Mediterranean and back to Italy or continue northwards towards Britain? Lee-Harwood continues,

> After some time it raised anchor, left Cadiz and headed north, which was when Andrew knew something was up. He kept this secret of course: he wasn't going to share it till he wanted to. The ship passed Brest in France then made a cut across towards the south-west of England. At this point Andrew blew the story.
>
> It was August, there was a lot of media space to fill. The *Karin B* was front-page news in every national paper for at least a week. There were

> endless 'SHIP OF DOOM' stories; there's something about a vagrant ship with a dangerous load that has a certain grisly romance to it, like the Flying Dutchman.

Regulations governing waste imports into Britain were anything but stringent at the time. The Government had declined to implement a key area of EC legislation that might have raised safeguards against such imports. Yet even the feeble national legislation that existed offered no harbour for the *Karin B*. It required exact manifests of any shipborne wastes to be filed and approved before they could come ashore for disposal.
At this stage, FoE's spot-check at Koko had been the only notifiable technical analysis of the cargo and it was in no sense an exact manifest, merely the results of a few dip samples from some of the barrels. So it was clear that the cargo wasn't acceptable under the rules.

> As the story developed, journalists quizzed the authorities on the regulations governing hazardous waste imports into Britain: that's when the whole can of worms was opened and spilled over the table. Within two weeks flat they'd implemented the European legislation that they had avoided for years and turned the ship away.
> The ship then sailed back to Brest, where it was excluded from entering the port, and ended up returning south. It tried to dock in Livorno, Italy, where it caused riots in the streets and pitched battles between the Carabinieri and local people throwing petrol bombs. Eventually it was accepted at Livorno and moored there. The cargo was taken off by officials wearing protective outfits and breathing apparatus.

The boat was sealed under an airtight canopy while the cargo was completely repackaged, itemised and put into safe storage. A key objective of the campaign was to ensure that countries which produced such wastes should dispose of them within their own territories. At the invitation of FoE Italy, Secrett travelled to Rome to help the group persuade the Italian Government and the European Community as a whole to accept their responsibility for disposing of the toxic waste safely in Europe. It worked, and in the face of a massive public outcry and Opposition pressure, the Italian Government reversed its position and agreed to look after the cargo.
An ironic twist in the *Karin B* tale was that the waste was ultimately transported back to Britain for disposal at the start of the 1990s. It was incinerated in Hampshire. Yet the story did not draw to a conclusion even there. Nobody ever found out who owned the *Karin B* and tales flew around that a boat of that name had been sunk off the Turkish or Syrian coast with another load of waste. According to Blake Lee-Harwood: 'When you enter the murky world of flag-of-convenience shipping you realise that all sorts of stuff goes on. It turned out there were probably *two* boats called the *Karin B*. One of them was sunk, presumably deliberately. But both were registered as the same boat. Work that one out.' Strange things were evidently afoot but FoE had by now withdrawn from this lengthy endgame. Says Lee-Harwood,

> It was out of our reach, we kept our focus on Britain, West Africa and that connection, that was enough. The Koko story is still one of the most authentic and best-documented cases of chemicals dumping in the Third World. Any article on the subject still uses Charles' Koko photos. It was also our first anti-pollution campaign in a developing country and, to be honest, I'm not sure we really knew how to deal with it in those terms. Though enormously well-intentioned where North–South issues were concerned, our understanding of development issues still had far to go.
>
> As far as Andrew and I were concerned, we saw something was wrong and just wanted to help. And we also learned a hell of a lot about toxic wastes. It's the kind of campaign that I guess you might think was more like a Greenpeace thing but it wasn't. All credit to Andrew, really. Many other people would either have been intimidated by the scale of it or said: hell, Nigeria's much too far away, we can't do anything about it. But Andrew had the guts and vision to just drop everything we were doing and say: yes, this is the biggest story this year. All sorts of positive things flowed from that. It was one of Andrew's great triumphal moments.

One of those 'positive things' was a full-blown Water, Waste and Toxics Campaign which involved Lees, Lee-Harwood, Mary Taylor and several others. Its broad base encompassed pesticides, incineration and municipal waste recycling as well as water issues. From about October 1988, however, drinking-water was the keynote concern. Lees had realised that EC Directives governing drinking-water standards constituted part of national law. Even though it was left to individual governments to set and implement standards ad hoc, it should be possible to determine current levels of contamination in drinking water, measure it against the agreed EC standards and, where there were failures, logically argue that a law signed up to by Britain had been broken.

The campaign team combed through records and asked the Water Authorities for all their drinking-water analysis data. The industry was reluctant to comply. A lot of the small water companies (unlike the big suppliers, the Water Authorities) were already private concerns, set up in Victorian times simply to supply drinking water to cities or towns rather than deal with sewage and wastewater. Many refused to help, or tried to be awkward, but persistence and pressure won through. The regional Water Authorities were still public bodies and found it hard to avoid handing over the information the campaigners requested, including data on pesticide, nitrate and lead levels in domestic tapwater.

Several enormous data-processing exercises ensued which involved dozens of volunteers and staff, many of whom were learning to use databases – another Lees enthusiasm – for the first time. A series of reports emerged on places where drinking-water quality was failing to match up to legal standards. Ultimately, the campaign team could show that almost 14 million people had been drinking water in British households containing illegal levels of pesticide residues. In 1989, veteran environment correspondent Geoffrey Lean wrote an *Observer* colour supplement cover feature based on the team's work.

It was splashed over four and a half pages with maps showing where drinking water had failed to match up to the standards. The switchboard at Underwood Street was jammed. There were stinging rebukes from the industry and furore in Parliament over what the revelations meant to the privatisation plans.

In the long, hot summer of 1989, sewage got the same going-over. More than 1,000 applications had been filed with Her Majesty's Inspectorate of Pollution for relaxations of water quality standards at individual sewage treatment works. It had become obvious that water companies would face heavy liabilities if they failed to meet the existing legal standards after privatisation. The Government's solution was to relax the standards. FoE argued that by putting would-be investors first, the Government was playing fast-and-loose with public health and nature by letting quality standards slip in drinking water and in the rivers where treatment works had waste outfalls. Mary Taylor says,

> In some instances, companies had not really been attempting to achieve the EC standards, despite agreement to the drinking-water law in 1980. We used hard facts and the law to show why proper compliance was important. That happened over nitrates, the Government had to concede that the figures couldn't be glossed over by presenting three-month averages or assurances that 160 per cent of the legal standard would be okay, which is how they were interpreting the legal limit on nitrates in drinking water.

FoE ran backup publicity campaigns inviting anybody whose drinking water was substandard to make a formal complaint to the European Commission, supplying pre-printed forms to fill in and post off to Brussels. The Commissioners were greatly impressed by the resulting flood of mail and in the end prosecuted the British Government over nitrates. The Government was found guilty.

The Government was also taken to the High Court by FoE over pesticides in drinking water. Though action on this score was slow in coming, partial bans on two of the most common culprits were set in place eventually, and timetables for compliance with the European standards speeded up.

Other aspects of waste disposal and 'management' also came under scrutiny. The big topic was landfill garbage dumps where there was a threat of groundwater pollution if toxic or potentially hazardous chemicals leached into watercourses. The pages of the *Observer* colour magazine and the friendly offices of Geoffrey Lean again supplied the medium; the message was a feature which listed and mapped some 1,500 risky sites. Says Taylor,

> We got more than 10,000 letters from people requesting information about Toxic Tips in their area. I think it must have been one of the biggest

Figure 20 Halton FoE takes samples of river water from the Mersey for pollution testing, 1991 © Chris Thomond

> responses from the public to any campaign message. We also got lawyers' letters threatening legal action from people who said their land had been blighted because of what we'd published. None of them came to anything but the fuss was enormous.

Though these were all quite complex issues, Lees was almost always able to create a drama out of them. By all accounts, working on these coups was no picnic. According to Taylor, fifteen-hour days were almost normal. 'We just kept going,' she says. 'It was pretty manic, but it paid off. We felt we were actually having an impact on the world and exposing some real scandals.' Adds Blake Lee-Harwood:

> It was fun to do. Very exhilarating. As members of his team, we were dedicated to the campaign and loyal to Andrew. It was essentially a case of him having the Big Ideas and others figuring out how to achieve them, including in my own case some of the media contact work. Everyone who took part feels proud to have been part of it. But it was always Andrew's inspiration at the bottom of it.

Tragically, Lees' life was cut short at the very end of 1994, at the end of a tour he had just completed in Madagascar. He had gone to check out potentially destructive mineral extraction activities by Rio Tinto Zinc on the southern coast of the island. He hoped to persuade the authorities on the island that the national exchequer and local people stood to gain more from promoting sustainable tourism in the mining area, whereas mining would only bring in temporary rewards that would vanish once the deposits were mined out. He also saw tourism as a way of safeguarding the area's many unique plant and animal species, for these were the attraction that would bring in the nature tourists.

Andrew died of heat stroke, weakened by a virus infection he had contracted towards the end of his visit. He had not taken rehydration solution out into the forest with him, nor the ingredients needed to make it. But this was characteristic of Lees' lack of heed for his own comfort when on the trail of the wild, and his total absorption in the task in hand, whether that happened to be botanising, campaigning for safeguards over his beloved Norfolk marshes, or crunching numbers on an overheated computer for a hard-fact exposé of would-be polluters.

Remembering Lees at a memorial service in London some months later, journalist Geoffrey Lean conjured up an affectionate picture of Lees in commando mode, gatecrashing a ministerial delegation that had come on a tour of inspection to Lees' marshland habitat. Lees had specifically been excluded from the party, though an acknowleged expert on the area. His outspoken reputation had preceded him and the visiting mandarins wanted no awkward questions to give accompanying newspaper reporters the wrong idea. But in the middle of the field visit, Lees came squelching undetected up the ditches and drains he knew so intimately and ambushed them in his waders, determined to give the Minister a piece of his mind, whether invited or not.

This piratical and fearless figure had been unflinchingly loyal to FoE's ideals, an inspired and inspiring advocate for the environment. At different times he had enraged practically each and every colleague he worked with. But they all felt a unique dash of colour had gone missing from their lives.

Working with Lees wasn't always easy. Roger Higman voices a reservation that as time went by would be echoed by other colleagues:

> Andrew was very interested in getting us publicity, branding and selling the organisation. And he was very good at the media. I learned an awful lot from him, he trained up a whole generation of campaigners. But – and nobody's perfect – his way of working was very wearing and his territoriality prevented him from achieving the fullness of things he did for the organisation.

Jonathon Porritt is equally forthright on the subject of Andrew Lees. He speaks of him as

> a great friend and a great opponent or sparring partner. Every single year I worked with Friends of the Earth I spent a lot of time debating with Andrew on the direction the organisation should go. And now and then saying: piss off Andrew, we're going to do this and I don't give a damn whether you like it or not. At times it didn't make me popular with him or his loyalists.
>
> The most articulate, powerful voices in Friends of the Earth have always been the key campaigners. It's part of the strength of the organisation. But it's also part of its inflexibility because once you invest that degree of power in such entrenched baronies, then the barons will sometimes make bloody sure that you don't move as fast as you need. And the barons have been a very significant obstacle to the kind of flexibility that's now needed, that lightness of touch I think the future will require. That's not a criticism of any individual. They've got a right to get stuck in there.

Porritt himself was evidently also getting stuck in to his leadership role at FoE. Many now feel his high profile in the media, his celebrity connections and his convincing advocacy for the green approach to life as a workable and realistic resource for problem-solving in modern politics, economics and civil society, did more to clinch the group's success during the mid to late 1980s than any other single factor.

Porritt also developed the practical solutions side of FoE's campaigning. The staff had always researched and promoted solutions in an academic way, but actual practical environmental action had largely been left to the local groups, working with little concrete support from the centre. Porritt spearheaded a number of national projects that helped to test solutions in the real world.

Taking advantage of funding available for practical environmental projects through the Government's UK2000 scheme, in 1988 he set up a Projects Unit within FoE. The Unit supported community-based recycling projects, ran seminars and conferences and published books to promote recycling and traffic reduction. Two projects, the Community Recycling Network and Paper Round, a company which collects waste paper from offices in London, were eventually launched as independent operations and still flourish today. But the biggest and ultimately most influential project was Recycling City, which ran from 1989 to 1992.

This ambitious initiative involved sometimes fragile partnerships between environmentalists, industry and local government in four areas of the UK. BT donated substantial funds to enable local authorities to test different approaches to collecting recyclable waste. The project aimed to investigate the obstacles to and opportunities for greater local authority recycling. In doing so it raised wider issues of waste management, and demonstrated that house-to-house collections for sorted recyclable rubbish were a serious proposition and not a green pipedream. Despite initial hostility from the campaigners working on waste issues, Recycling City also showed those inside FoE that the affirmative style of campaigning which excited Porritt, as a counter to the more adversarial and contentious stance favoured by Rose and others, could be made to work to equally emphatic effect.

Another such initiative was the *Environmental Charter for Local Government*, published in November 1989. The Charter was drawn up partly in response to growing demand, but it also aimed to build up local authority staff and politicians as a constituency for FoE. Building on the tradition started with the bestselling *Bicycle Planning Book* of 1978, the Charter combined policy demands with detailed background information. Based on a fifteen-point declaration of commitment, it was supported by 193 separate policy recommendations, with practical implementation advice, which could be adopted either individually or as a package. Advance consultation with friendly local authority contacts helped get the language right for councils. Some twenty authorities adopted the declaration as their own policy statement.

Duncan McLaren, the Research Officer given the task of compiling the recommendations in 1989 and now head of FoE's Sustainable Development Research Unit, recalls: 'We knew the Charter was going to be a success when, weeks before publication, enquiries started to come in from councils – including one Environment Officer who had a photocopy of the consultation draft. But it had been copied so many times he couldn't actually read it.'

The successful formula has since been repeated in books including the *Recycling Officer's Handbook* (1991), *Less Traffic Better Towns* (1992) and *Planning for the Planet* and the *Climate Resolution* (1994). Together, they have led to a real culture change in local authorities, not least because they helped local FoE activists to get key jobs in authorities across the country. Duncan McLaren says,

> In the last six years, we have seen a shift in local politics from the environment being largely a marginal issue, to most authorities having policies, many closely based on the Charter. And today, as local authorities face up

> to their Earth Summit mandate – to establish their own Local Agenda 21, the value of the Charter initiative can be seen clearly in the fact that they see FoE nationally and locally as an important contributor to their discussions and plans.

Porritt's desire to play a figurehead role in campaigns that specially interested him occasionally embarrassed the senior campaigners. Chris Rose was running the Agriculture campaign when Porritt first became involved back in 1984. He feels the inexperience of the new Director as a frontline campaigner showed at the beginning:

> By the time he left Friends of the Earth he knew much better how to campaign. But I remember him wanting us to have an agriculture policy campaign. We said: you can't have an agriculture campaign unless it's about what's wrong with agriculture. He said we should have a campaign to push organic farming. Of course now in a way you could. But then the way pressure-group campaigning worked, you couldn't do that. You could campaign against pesticides or nitrates, something nasty.

He also cites an abortive intervention by Porritt in the issue of peat extraction by small-scale whisky distillers on Islay, in Scotland's Western Isles. The peat beds were important as feeding and breeding grounds for wild geese. 'His idea was to go up to Islay with David Bellamy and educate the islanders about what a bad idea it was to dig up peat and what a good idea it was to protect geese,' recalls Rose. 'They were practically lynched. They got stuck on the island and had to go into hiding in a friend's house.' Porritt and Bellamy were eventually escorted off the island by police, for their own safety.

Nevertheless, the Duich Moss campaign has gone down in the environment movement's history books. The Greenland white-fronted goose is one of Europe's rarest birds. Its winter roost site, Duich Moss on Islay, was supposedly protected under both UK and European wildlife law. The decision of the Secretary of State for Scotland, George Younger, to allow planning permission to cut peat for flavouring the distinctively smoky Islay malt whiskies was a conservation scandal that could not be ignored. However Porritt's decision to send in outsiders to blockade the bulldozers was undoubtedly foolhardy, and the headline on the leaflets produced for a public meeting on Islay, 'Whisky or Wildlife' was strategic suicide. As a local shouted out at the meeting, 'I will tell you Friends of the Earth, you have picked the wrong time, the wrong place, and the wrong people.' A fortnight later FoE sought to explain its case better in a leaflet more wisely titled, 'Work, Whisky and Wildlife'.

The bulldozer blockade may have only lasted three days, but activity continued off the island. A complaint to the European Commission that the UK had ignored the EC Wild Birds Directive resulted in the first ever visit of Commission lawyers to a member country to investigate an alleged breach of wildlife law. The end result was that the Guinness-owned Scottish Malt Distillers abandoned their plans to dig up Duich Moss, getting peat from another site. The Greenland white-front's roost may have been saved, but the Nature Conservancy Council, who started the whole affair by challenging the original planning application, and whose local officers encouraged FoE's involvement, ended up with their feathers severely singed. The NCC is no more. It was split into three separate agencies, one each for England, Scotland, and Wales, by the Secretary of State for the Environment, Nicholas Ridley, for the declared reason of making nature conservation 'more sensitive to local sensibilities'. Conservationists see it rather as revenge for a campaign that resulted in the political embarrassment of having an EC lawyer interfere in UK domestic business.

Further ramifications of the events on Islay were felt when peatbogs once again became a campaign target in the 1990s. Following up reports from the Yorkshire local groups, FoE shone the spotlight on commercial peat extraction, and in particular the activities of horticulture and pharmaceuticals giant, Fisons, which was destroying important peatbogs at Thorne and Hatfield Moors, near Doncaster. In alliance with other groups, notably Plantlife, RSPB, and the Wildlife Trusts, FoE launched the Peatlands Campaign. The peat industry didn't know what had hit it. Attacks came from all sides – shareholder questions at Fisons' AGM, boycott stickers on peat products in gardening centres, debates in Parliament, giant banners on the moors and at the Chelsea Flower Show telling them unequivocally to 'Bog Off'.

But in 1992, just when victory seemed in sight, English Nature intervened with a rescue package that allowed Fisons to dig up nearly a third of its 8,000 acres of prime peatbog, all of it designated as Sites of Special Scientific Interest. Robin Maynard, FoE's Countryside Campaigner at the time, still feels angry at this dereliction of its duty.

> We had Fisons on the ropes, peat sales were falling, B&Q had just announced that it wouldn't stock any of their peat products, the alternative peat-free composts were coming on stream. And then the English Nature suits waltzed in to give Fisons a seal of approval to continue destroying some of our best wildlife sites. Ridley had clearly achieved his aim of gutting the NCC after Duich Moss.

For all this, by championing selected causes, Porritt had shown that FoE could not only propose solutions to environmental ills, it could actually join with its local groups and other partners in the community to test those solutions out at community level. Charles Secrett believes that such ventures furthered the Act Locally tradition of FoE in a vital new way. 'They made people see that the environmental agenda is not just a pipedream, that power does rest with the people.'

14

Twin peaks

I sat up in my hammock and gazed out across the plain at this thrilling and majestic sight. I suppose that Billali noticed it, for he brought his litter alongside. 'Behold the house of She-who-must-be-obeyed!' he said. 'Had ever a Queen such a throne before?'

'It is wonderful, my father,' I answered. 'But how do we enter? Those cliffs look hard to climb.'

'Thou shalt see, my Baboon. Look now at the plain below us. What thinkest thou that it is? Thou art a wise man. Come, tell me.' I looked, and saw what appeared to be the line of a roadway running straight towards the base of the mountain.

– H. Rider Haggard, *She*, 1887

While the pendulum was swinging between different points of view on the organisation's main role, one campaign area, energy, remained a steadfast law unto itself. In the nuclear issues arena, the tussle over radiation risk levels had been won but Pad Green (who had joined the staff as Radiation Campaigner in 1990, having completed his PhD) and Simon Roberts, who had succeeded Stewart Boyle as Senior Energy Campaigner, had no illusions about the time it was likely to take to bring about change in favour of safe energy options as a result. Green remarks,

> The nuclear industry is probably the most powerful vested interest in the world, or it certainly has been. Nuclear power permeated British establishment thinking on all sides for decades: a Labour Government approved THORP and pushed nuclear power in the sixties, the 'white heat' and all that. Then there was Margaret Thatcher. You couldn't get much more of a 'white heat' junkie than Maggie. Her belief in nuclear power also went with her belief in smashing the coal miners. And there was also the pure Free Market ideology, the idea that the City was always right.
>
> It's all very well telling somebody that something is wrong, but you've got to give them a way out. And you've got to recognise that you don't always get all you want at once. In trying to change things, you have to learn to put yourself inside the other side's head. Otherwise how are you going to change their thinking? Back in the 1950s nuclear power offered a

> good deal of promise: as it was, it couldn't deliver it without prohibitive risks and costs. We could prove that, but then to expect that the Government would pull the rug from under the feet of the nuclear industry was irrational. It was obvious that they were going to give it another try. Which is what they did.

The next 'try' turned out to be the global warming issue, then shaping into a key international debating point, and an apparently heaven-sent selling point for a beleaguered nuclear industry. The Intergovernmental Panel on Climate Change (IPCC), a group of scientists and policy analysts formed under UN auspices, had begun to establish a worldwide consensus that carbon dioxide levels in the atmosphere were climbing perceptibly as a result of industrial emissions, most notably the constant welter of exhausts from power plants burning fossil-based fuels.

It didn't sound much of a percentage rise, but it was enough in the view of many top climatologists to threaten a significant and accelerating rise in air temperatures around the world. The Sun's energy, reflected from the Earth in the form of heat, was increasingly being prevented from escaping back into space owing to the heat-retaining properties of carbon dioxide and other greenhouse gases accumulating in the upper atmosphere.

Unless greenhouse gas production was regulated, the IPCC warned, global warming, sea-level rise and climate change could result. Low-lying coastal centres of human population could be engulfed and agricultural production blighted as deserts spread into areas presently seen as bread-baskets.

Not long after the outset of the public inquiry into a proposed nuclear power station at Hinkley in Somerset, British Nuclear Fuels Ltd launched a million-pound public information campaign in the national press, proclaiming nuclear energy as the 'clean and green' energy option that would, in effect, save the planet.

The strength of the case Friends of the Earth put up against this sales talk was that it answered the climate arguments as well as stressing the risks of radiation. Pad Green remembers:

> We thought long and hard about what was the most cost-effective way to make a difference. The vulnerable point we pressed on was that the risk of radiation was being undersold. We reckoned that raising public awareness of the radiation risk would have several knock-on effects. It would drive pressure for safety standards – and form allegiances with people within the industry and make them more concerned about what they were doing, which in the longer term is quite an important thing. Also if they have to improve safety standards the costs go up, and then costs begin to reflect the real risk.
>
> We were essentially trying to get the industry into a corner over costs. There's an idea going round that it was economics rather than environmental protest that finished off nuclear power. But what is it that has made nuclear power uneconomic? You can't separate the environment from economic considerations and environmental pressure groups made sure that market costs reflected the real risks.

FoE's work on global warming was not being simply driven by the nuclear industry, however. In 1987 Adam Markham, then a Pollution Campaigner, had commissioned *The Heat Trap*, a comprehensive report on the current scientific awareness of climate change, and this was published to loud acclaim in 1988. The organisation was widely seen as being a leading opinion-former on an issue that most people deeply feared but few really understood. More significantly, what had started out as a scientific debate quickly became a media rollercoaster of breathtaking dimensions.

Fiona Weir, head of the atmospheric pollution campaign from 1988 to 1995, recalls the orgy of press coverage:

> 1988 was the year that global warming really happened for the public. There had been a whole series of natural disasters round the world, and a big summer drought in the UK, and suddenly there was a theory that seemed to make sense of it all. And it wasn't just the usual doomsters who were saying it but the world's top scientists. Even Thatcher chose 1988 to reveal her green credentials.

Not surprisingly, the heady mix of news-hungry journalists and complex theoretical science ensured that green groups didn't always get top marks for their facts. And while FoE occasionally got some of the technical details wrong, there were other less scrupulous groups who didn't hesitate to publish maps showing Birmingham as an island and Blackpool under water.

As an information-hungry public became caught in the crossfire of contradictory facts being fired by industry, Government and the greens, it was obvious that the global warming debate was going to be a test case of the environmental movement's credibility. And if a lack of factual consensus was proving confusing, the media was going through a bout of almost total schizophrenia. Fiona Weir remembers that

> the press, especially the tabloids, couldn't make up their minds about whether global warming was a good or bad thing. One day the front pages would threaten instant incineration for their readers and two days later they would be extolling the joys of going to work by cabin cruiser after sea level rise and getting a sun tan in Scotland.
>
> We didn't always get it right either. Our first local group action involved people dressing up in snorkels and flippers on the high street. It seemed witty at the time but in retrospect was really inappropriate. We still haven't really succeeded in driving home the truth about global warming to a wide audience.

By 1990 the debate had begun to settle down but the omens for planetary disruption had become yet more threatening. The second World Climate Conference in Geneva brought together the IPCC, governments, industry and scientists to discuss likely changes in climate and culminated in the scientists calling for substantial cuts in the so-called greenhouse gases. It was this conference that led directly to the signing of the Climate Convention at UNCED in Rio and the commitment by world governments to 'aim to return' their emissions of greenhouse gases to 1990 levels.

However, international conventions notwithstanding, there has been precious little action on climate change. Sir John Houghton, chairman of the IPCC, was still warning in 1996 of the flooding of densely populated regions within a century, the migration of millions of environmental refugees from sea-flooded deltas or drought-stricken fields and the ever-present threat of warfare over dwindling natural resources.

But despite the gloom, Fiona Weir retains some optimism about the future of the planet.

> It's true that there's not been much action so far but there has been a huge leap in understanding of the issues and the first, albeit token, attempt at international action through the UNCED Climate Convention. The political will necessary to tackle the problem will ultimately develop in response to mounting evidence of disaster – the real question is will it be too late?
>
> And what you have to remember about climate change, is that it differs from other pollution threats in its solutions. Most pollution problems have a technological solution – we can replace bad chemicals with good ones, cut sulphur emissions from coal-fired power stations and so on, but there are no such solutions for climate change. The only way we can tackle it is through adapting our lifestyles – and that's a challenge the industrialised world has never had to face before.

And the lessons for the environment movement? Weir is clear about where the debate left FoE.

> The whole global warming debate gave a huge boost to our credibility. Apart from a few minor errors early on, we got our facts right and were telling the public the truth about climate change long before the Government. The respect this won us has stayed with the green movement and opinions polls show that environmental pressure groups are still far more trusted as a source of objective information than Government or industry spokespeople.

The work of the IPCC on climate change grew partly from a related international debate on ozone layer depletion, related not least because the industrial chemicals implicated in the problem, the CFCs (or chlorofluorocarbons) also happen to be greenhouse gases. The UN Environment Programme had convened a series of global meetings in response to reports showing that a gaping hole had appeared in the ozone layer over the South Pole, bearing out predictions made many years earlier by James Lovelock.

Adam Markham, prescient as ever, had understood the scientific debate in the mid-1980s and saw that ozone depletion was going to become a major environmental issue (although interestingly, FoE had been sent a proposal for an ozone campaign in 1980, before Markham's time, but had rejected it). Consequently, Markham set to work in the summer of 1987 to run a campaign against ozone-destroying chemicals

Figure 21 FoE delivers a 'package of measures' to tackle overpackaging to the Department of the Environment, 1993 © Mark Mather/FoE

in fast-food packaging. He then compiled a positive list of CFC-free aerosol products, to be issued to shoppers in a bid to encourage ozone-friendly purchasing.

Meanwhile, on the international stage, the nature of the crisis was recognised by the signing of the Montreal Protocol which committed governments (mostly from the industrialised North) to 50 per cent reductions in CFCs and a freeze on halon gases.

As the dangers posed by CFCs became clearer, FoE's tactics changed. With the departure of Markham at the end of 1987 and the arrival of Weir, it was decided to drop the 'buy ozone-friendly' consumer campaign and threaten aerosol manufacturers with a stark choice: dump CFCs or face a boycott.

In February 1988, Weir sent the manufacturers' trade association a sample pack of the materials that FoE planned to distribute to the public; it made grim reading. Meticulous research had identified exactly which household names were using CFCs. Aerosol maufacturers saw a tidal wave of bad publicity poised to descend on their carefully nurtured public images.

The manufacturers didn't take long to answer FoE's challenge – exactly forty-eight hours – and capitulated with a promise to remove all CFCs within two years. This wasn't quite fast enough for FoE. Weir unleashed 200 local groups onto the high streets of England and Wales the next Saturday in a nationwide pavement campaign to shame the aerosol industry. The results were spectacular and the media went wild. Weir recalls,

> It was an astonishing time. The ozone issue was everywhere from *The Archers* to the Royal Family. I seemed to spend my whole time on TV and radio – the news, *Blue Peter*, you name it. And we formed bridges with groups that had never heard of us before. The Women's Institute even passed a resolution at their national conference.

With the aerosol issue almost won, and Markham's original food packaging campaign going in a similar direction, it was time to start picking off other industrial sectors. Weir's team of crack researchers, including an indefatigable volunteer, Mick O'Connell, combed through company information seeking out the ozone offenders and naming culprits. First the building industry was taken to task over CFCs in insulating foam then, in 1989, the campaign moved on to tackle industrial solvents while also mopping up other issues such as the disposal of old CFC-laden fridges.

In 1990 the Montreal Protocol was renegotiated in London and a complete phase-out of CFCs agreed along with the addition of several other ozone-depleting chemicals. The momentum appeared unstoppable and at Copenhagen in 1992 the dates for CFC phase-out were brought forward again. Not only that, the pesticide methyl bromide was added to the list along with the ozone-depleting HCFCs. The HCFCs had only just

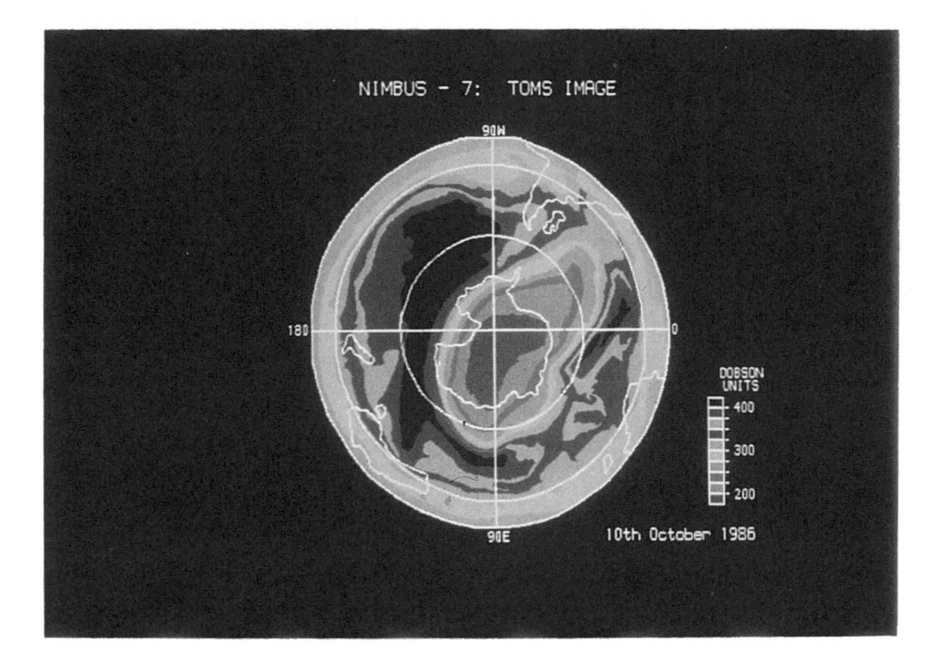

Figure 22 The Antarctic ozone hole measured by the Total Ozone Mapping Spectrometer, 1986 © DRA/Still Pictures

been launched into the marketplace when they were placed under strict controls and eventually banned.

The Copenhagen negotiations meant that all known ozone-depleting chemicals were now controlled by the Montreal Protocol and that the issue had effectively been won by the green lobby. Fiona Weir thinks that FoE learned some valuable new tricks from the ozone debate.

> It was a classic pressure-group campaign and we learned a lot of lessons: we learned that consumer power, properly harnessed, was a big stick to wave at industry; and we learned that meticulous research of technical sectors could really lift the lid on business and frighten half of them witless.
>
> Perhaps most importantly, we managed to build bridges to constituencies that we hadn't reached before, people who didn't think of themselves as environmentalists but who shared our views and values on a particular issue. And the long-term benefits from that kind of alliance building are still with us.

Towards the end of the 1980s, FoE was riding a wave of official approbation. Thatcher's 1988 speech had given right-wing newspaper editors the cue they needed to take environmentalism out of the radical ghetto and place it on the public agenda, happily bestowing the mantle of green guru upon old Etonian Jonathon Porritt in the process. Even the *Daily Telegraph* gave Porritt, and the youthful readership the marketing men felt he represented, a generous crack of the whip in terms of inside-page column inches.

But the momentum of FoE's campaigns seemed to some onlookers to slacken in this unwonted atmosphere of official approval: when the Government set a new challenge for local authorities to recycle at least 25 per cent of municipal wastes, for example, FoE and its local groups found themselves in instant demand as advisers. Even so, discontent among local groups was surfacing once more. Some felt put out by the national organisation's income-generating activities, which they saw as siphoning off local support and courting the armchair donor rather than the local activist. Nor did they see benefits coming their way from the increased spending-power the fund-raising bestowed. Others felt the organisation was becoming ineffective as an agent of change in relation to government and industry.

Still others felt excluded from the campaign side of things. Andrew Lees eventually took over from Charles Secrett as Campaigns Coordinator when the latter left for Brazil in 1989. Lees had favourite local groups with whom he worked well on occasion. 'If you were a local group that he thought was competent you'd have a whale of a time with him,' says Iris Webb. 'He operated a kind of mediaeval patronage system. If he thought you were good, he'd give you anything.' The also-rans, however, were liable to feel left out in the cold and some began to size up ways to take the situation in hand.

Val Stevens saw both sides of this growing divide. As a prominent founder member of one of the most successful local groups in the country, FoE Birmingham, she had been intimately involved since 1974 in that group's early struggles to establish a foothold in the city, and its gradual evolution into a well-to-do and efficient campaigning force.

A lucky financial break had set the group on course when the *Sunday Times* featured its envelope re-use stickers in a prominent consumer feature. Orders had flooded in from all over the country and the proceeds had helped buy a lease on former warehouse premises in the city's Digbeth area. A wholefood wholesaler took on a share of the lease and supplied an in-house wholefood shop. Other like-minded outfits rented space in the building, which in time became an alternative resource centre held in high esteem in the local community. By juggling with job-creation programmes in the 1980s, the group had grown to become a major source of work for more than 100 locals, who were mostly engaged to install insulation in the homes of pensioners and other disadvantaged groups.

Birmingham's enterprise became, says Stevens, a phenomenon of more than local note. In 1981, the national FoE office was given the chance to feature in a BBC2 *Open Door* programme. Stevens remembers,

> They said: we've nothing to show except a lot of cluttered desks. Why not go and film Birmingham, they've got so many things going on there. We did an entire half-hour programme, showing the vans going out to insulate houses, the shop, the paper sales. At that time we didn't have a café but we'd organised a canteen. It was always somebody's turn to do lunch, quite often cooking for forty or forty-five people, the queue stretched out of sight.

So it was always a very convivial place, always buzzing with chat and activity. It made an impressive bit of television.

Stevens herself was going out to give talks and run courses for schools, churches, charities and other organisations. For years, she says, she did probably three talks a week somewhere in the city. She was also Chair of a Save Public Transport Campaign, and of an organisation called the Green Ban Movement, recently started up in Australia by Jack Munday, whose aim was to get trade unionists to block destructive mining projects and inappropriate city-centre developments. Says Stevens,

> It seemed to us that what Munday had achieved was to bridge this terrible gap between middle class environmentalists and working people. Many of them had deep suspicions that this movement was going to ruin all their jobs. What I did here, with the help of sympathetic people in the union movement, was invite union members to come to debates about nuclear energy. Though Green Ban was mainly focused on city-centre development, I thought if we can get them all together on this and on public transport, we've got a foot in the door. So we were talking about safe energy alternatives, saying let's not go down the nuclear road.
>
> In the 1980s of course it all started going yuk, there was unemployment, everyone started getting nervous, union membership went down, they couldn't take on new struggles.

In Jonathon Porritt's time Val Stevens became closely involved in the relationship with the national organisation. She had known Porritt for several years and he had invited her to strategy meetings with the Board before she was elected as a Board member in 1987 on the strength of her activist record. She recalls,

> It was a heyday. The money was beginning to roll in because of the big mid-eighties resurgence of public interest. I remember at one strategy meeting a £6 million budget was forecast. The staff payroll in London rose to about 120 and the organisation was growing so fast that we were appointing new people every other week.
>
> I became Chair of Personnel when FoE was setting up a Personnel Department. That was when I became so conscious that all this massive funding pouring in from the public and this huge growth of staff were all in the centre. People in Birmingham were still working for about £3 an hour and had no holiday entitlement, pension rights, maternity leave, none of the things being bargained in London. The local groups department at national office had always been run by one person. In my time it was Chris Church. With a bit of backup he managed to service 250 groups round the country. After the expansion the local groups department grew to five people.
>
> They seemed to spend so much time trying to sort themselves out and develop strategies there was almost a feeling that this new team was less effective than one guy running his own show. Another department that grew enormously was fund-raising, that was how all this extra money came in. It grew from one person to ten. Then came the corporate image thing. We weren't going to be scruffy ex-hippies any more, we were going to have a

> proper image and all our publications were going to be orchestrated, it was all going to be run by professionals who'd give us this professional image and Do Things Right. I think that created quite a lot of tension.
>
> We felt we'd already demonstrated our credibility and usefulness and commitment to the community, and our professionalism, too. Some of the newcomers at the national office wanted to tell us how to do things, what logo to use, what kind of banner to have. Previously, any group that had a good idea, like a teachers' pack on whales, could then sell it to all the other groups. But now we weren't left in it, it all had to come from the centre, where there were professional people who knew how to Do It Properly. I resented that and I know other people did. I fought like hell against it on the Board.

Whilst the 'professionals' at head office may sometimes have seemed unsympathetic or cavalier towards the local groups, they did make changes that sustained FoE in the leaner years which followed. For example, professionalism in the use of direct mail led to enormous growth in membership and, crucially, in the number of members donating through standing orders. Between 1988 and 1990, national membership jumped from 38,000 to over 200,000 and the income raised through standing orders multiplied by an order of six. This gave the organisation sufficient financial security to come through the recession with far less contraction than many rivals. The appointment of professional design and publishing staff also enabled the organisation to communicate its message more widely and in more sophisticated ways than would otherwise have been possible. The in-housing of essential computing services saved money and enabled far greater flexibility in the use of electronic information.

Eventually, funds began to flow more freely to the regions. In 1990 two Regional Officers were appointed to support the local groups. Following the success of these appointments, the team was expanded and now stands at eight. Just as the first Regional Officers took up their posts, FoE Birmingham characteristically jumped the gun by appointing and funding its own campaigner, Elaine Gilligan, to chase key issues locally along lines determined by Underwood Street. Gilligan would later become the Regional Officer for the Midlands and now works in the local campaigns department at the national office. Val Stevens says,

> When she was a Campaigns Officer here, Elaine began to feel strongly that though we appreciated the importance of the main campaigns, rainforests, countryside, land use and so on, she wanted to move into the area of local campaigning on the roads issue round Birmingham. Elaine wanted to see local action on issues that mattered to people, to show people that Birmingham FoE campaigned to improve their lot and not just on issues that could seem remote. Tackling local transport issues was an obvious choice.

In April 1991, Birmingham City Council announced plans to upgrade two link roads between the M40 and the city centre. Both went through thriving local communities with small businesses, local amenities and shops. Gilligan helped set up joint meetings with the Asian Traders' Association, residents' groups and other key local interest groups in the road scheme's neighbourhood. The response, says Val Stevens, was electric.

> In a way she was doing what I'd wanted to do in the seventies with the unions, but with real communities rather than other organisations, people who lived round a section of this road and felt concern over what was going to happen to their surroundings. And that really got things going.

For the next five months Birmingham FoE campaigned within a coalition of community groups that finally saw the Council back down in the face of pressure from an unprecedented cross-section of outraged citizens' groups. Birmingham FoE was now set to be a key player in fighting plans for an orbital motorway in the Midlands.

Within the West Midlands region as a whole, plans were afoot for an orbital motorway composed of a Birmingham Northern Relief Route, a Western Orbital Motorway and a widened stretch of the M42. The transport activists started talking to other groups in the region in Coventry, Solihull, Dudley, Walsall and Wolverhampton. That began to bring in more rural groups from as far afield as Nuneaton, Cannock, Stafford, Wyre Forest (Kidderminster) and Bromsgrove. Says Chris Crean, an activist for Birmingham FoE:

> We were building a big network of urban and rural groups to join together when necessary to fight these orbital schemes. We set up what we called the West Midlands Transport Campaign to hold meetings and keep everyone informed on developments. We also found there were local community groups fighting these schemes. Some of them probably felt Friends of the Earth was a bit too far out for them. But we started to form alliances with them along the proposed routes of the schemes. We said, form your action group and sign up to one rule we'd set which was: we object to this road in its entirety. Anybody who'd do that could then join the alliance and share our information and media coverage.

The campaigning activities of local groups had not gone entirely unnoticed at Underwood Street, however. Chris Crean explains,

> One thing we've noticed in the last few years is much greater support for local groups and their activists working at a local level. A fund has been set up to support campaigns which local groups have identified as local priorities. In addition, the staff now prioritises three or four key campaigns each year as organisation-wide priorities. The whole organisation, staff and local groups, work on targeted issues at the same time. Since the inception of this way of working in 1994 the roads campaign has been chosen. This has resulted in the West Midland Transport Campaign receiving both campaign and administrative support from staff at Underwood Street and the Midlands regional office.

Jonathon Porritt had been planning to move on since 1988, to make an environmental TV series. His career in FoE had been a rollercoaster ride but he had proved himself equally able to cope with boom or bust. David Gee, a former campaigner who had since had a successful career in industrial relations, was appointed to succeed Porritt, with a period as Campaigns Director as a form of training. Porritt finally left in the summer of 1990.

At the very end of the 1980s came the collapse of the Berlin Wall, hardcore emblem of the Cold War that had provided so much of the psychological drive behind postwar concern for the environment. Perestroika had put paid to the threat of Mutually Assured Destruction that had hung over humankind's head for more than a generation. It was hard to see how professional doomwatchers were going to manage without its potent echo in their own oratory.

David Brower's Think Globally, Act Locally caravan was fast approaching decisive run-ins on the twin peaks of the Earth Summit and Twyford Down. Had its motivating drive been taken away before it got even halfway there?

15

The road from Rio

The first rule of intelligent meddling is to keep all the pieces.

– Aldo Leopold, *A Sand County Almanac*, 1949

Unfinished business was the order of the day as David Gee took over as national Director. Gee knew that Porritt's was a hard act to follow, many of his new colleagues were at loggerheads over policy and finances were shaky. But it was hoped that his know-how and flexibility as a professional conciliator and human resources manager would help put the organisation back on track. The Board made encouraging noises. They recognised that the unprecedented boomtime of the late 1980s had led to unsustainable growth and financial strain. The need now was to consolidate. Says Iris Webb,

> Boom and bust was almost endemic to FoE's financial situation, like population cycles in nature. The Board thought they wanted this calm manager type who'd allow everyone's talents to flourish. That would allow sustainable growth to happen and make the machinery work again. What I think they really wanted was a charismatic figure who'd stand up and do what Jonathon did, maintain a high public profile and handle the Andrew Lees types within. And that David never was, he wasn't expected to be. He was absolutely competent, had a degree in chemistry, knew how to campaign and how to manage. But that pressure-cooker culture in head office was something else, it really didn't suit him. I think it made his life pretty miserable.

In overall structure, the organisation looked sound enough. Most of the 250 local groups were working independently with growing confidence and effectiveness. Another plus was the central fund-raising support offered by Friends of the Earth Trust. Winning charitable funding meant tax breaks and more freedom for campaigners to buy in new research. By the early 1990s, the Trust had it down to a fine art, thanks to the dedication of a number of high-calibre trustees, including Godfrey Bradman, ace fundraiser Sherie Naidoo and trade union chief Clive Jenkins.

Meanwhile, the formerly dedicated research wing of the organisation, ERR, had become an entity apart. During the lean years of the early

1980s, it had no choice but to broaden its consultancy base beyond the intermittent pickings FoE could pass its way. The parental connection had atrophied.

Internationally, FoE's prospects were healthy. New national chapters were being founded around the world, especially in the global South and in the newly democratised countries of Eastern Europe and the former Soviet Union following perestroika. By 1992 the global membership profile of the organisation bore little resemblance to its original, largely white, middle class, Anglo-Saxon constituency, which now formed a numerical minority.

And yet, in common with other pro-environment groups everywhere, FoE was finding it hard to take stock of where in the world to go in the wake of the Earth Summit, against downbeat economic trends. Former FoE campaigner Colin Hines sums Rio up as 'a watershed in the sense that we no longer had to tell people there was a problem. We now moved into what the hell you do about it, and what governments do about it.'

In that perspective, Hines feels, Rio was a culmination of all the work that had been done in previous heydays of environmental concern. It enabled – or forced – green organisations to reconsider what they were doing. 'We spent years grabbing people by the lapels saying, look, this is important,' he notes. 'While at Rio people were saying okay, I know, so what do we do about it?' Hines thinks this shift towards a problem-solving mode has happened at a time that is anything but ripe:

> You're unfortunately looking at a period when, though we've won the argument, a fresh trend for deregulation is coming in. All the advances the environmental movement has made, particularly in the United States, are now being rolled back. Governments are saying, look guys, we agree but we can't afford these reforms. They're necessary for the future but not now. Don't introduce them or investors will relocate out of the country.

Hines feels sure the problem he describes is hitting not just environment but social concerns as well, suppressing a groundswell of acquiescence in increased taxation to meet agreed social or healthcare needs.

Many within the movement take a stronger line on the significance of the Earth Summit's effects. The clans had been summoned to Rio on the basis that climate change and other issues had become too big and too intractable to handle at national level. Only an orchestrated One World fix would do. The scale and cost of solutions would otherwise be prohibitive.

Yet the message governments had issued when it came to the crunch was that the money wasn't there anyway. In Agenda 21 they had in effect handed down a do-it-yourself blueprint to the grassroots with very little in the way of an accompanying commitment on their side to make new funds and resources available for implementation, or to fulfil their own part of the bargain on reforms at top level.

Where was the much-vaunted Peace Dividend many had expected to flow from the healing of Cold War wounds following perestroika? The arms race had absorbed the majority of development investment and research talent since World War II but it was surely history now. How, if not on environmental rescue, was the resulting bonus of material and human resources to be spent? Surely not on unbridled wealth creation and industrial expansion to protect the world's Haves against deepening recession at the expense, as usual, of the Have-Nots and of the natural world's security and stability?

Yet that was the way it looked to many who were trying to gauge the state of things in the aftermath of Rio. Perhaps part of the curious sense of a loss of momentum many felt was itself to do with the state of geopolitics after the fall of the Iron Curtain. Any idea that an Age of Reason was arriving had been quickly dispelled by the horrors of the Gulf War. But by and large the nightmare of nuclear holocaust had been driven from people's minds and with it perhaps a sense of urgency about bringing other life-threatening global ills such as climate change under control.

Apart from trying to make sense of the post-Rio agenda, there were more immediate concerns for FoE to get to grips with in the early 1990s. On the Energy front, as so often before, the way forward looked clearest and confidence was up to scratch.

The Government was talking about privatising the nuclear industry, so Pad Green and his team focused on that. The 'clean and green' image the industry was going to extravagant lengths to project was packaged more for consumption by the public than by would-be investors. Both FoE and Greenpeace were focusing their energies instead on convincing the City that the nuclear industry did not know how to manage its waste or decommission its reactors. Safety standards would, they argued, be compromised for sure and the industry's insurers would increase the premiums. Green recalls,

> We said, we know nuclear power's uneconomic, the public thinks it's unsafe. Industry's trying to convince the public it's okay but the Government's still trying to decide what to do. Then, the City turned on the Government and said, you're asking us to sign a blank cheque for *this*? We're not going to buy it.

The Government's response was to offer investors a protected share of the market, in the form of a subsidy for energy production by means other than fossil-based fuels, created by charging a 10 per cent Fossil Fuel levy on the energy bills of every UK household and business, a deal labelled the Non Fossil Fuel Obligation (NFFO).

The NFFO measure was ostensibly introduced to meet the nation's commitments to curbing greenhouse gas emissions, entered into under the Convention on Climate Change bargained at Rio. But its subtext was to lure investors into buying out the nuclear industry. The sweetener was calculated to reassure the City that more than enough revenues would accrue from the levy to cover liabilities in respect of radioactive waste disposal or reprocessing spent fuel, and of decommissioning reactors once their shelf-life was played out.

'Nuclear energy was simply more expensive so you had to give somebody a legal requirement to buy it', Green explains. The levy achieved that by subsidising nuclear power, thus removing any bar to using it on the grounds of cost. But the NFFO and its advertised function as a way of phasing out fossil-based fuel options had a double edge. Environmentalists saw an opportunity for wind power and other fledgling renewable energy industries to scale up to profitability at the levy's expense. Sympathetic civil servants concurred. A brave new world of energy alternatives came to life. Says Green, 'We saw it as a way to say, nuclear power is uneconomic so let's make it unnecessary in the long term. We don't need it if we get renewables off the ground.'

As the powers-that-be finally began to blow cold over nuclear energy, the roads issue became more and more a political hot potato. In the aftermath of the Twyford Down and M11 protests, the roadbuilding ambitions of the Department of Transport were scaled down, after a succession of hurried climbdowns from nearly all existing major road proposals, to a handful of big infrastructure projects and one highly contentious scheme for a bypass around the town of Newbury.

From marshalling dissent against specific road schemes, a territory now occupied by a plethora of highly organised single-issue groups, FoE's Transport Campaign began to take a more general turn by the end of 1992: a head-on tilt against what Margaret Thatcher had once termed the Great Car Culture. It grew increasingly evident throughout the early 1990s that British public opinion was ready for a fundamental re-think on this score.

A report by Martin Linton in the *Guardian* in August 1995, for instance, described how protesters on bikes campaigning under the banner of one of the emergent single-issue groups, Reclaim the Streets, had brought chaos to the capital by blocking rush-hour traffic in Greenwich and creating car-free zones in several town and city centres during the same fortnight in summer. The newspaper commissioned a poll from ICM to gauge the reactions of bystanders, including motorists who were being roadblocked, to this protest.

The poll revealed, wrote Linton, that 'most people, even those who have cars, agree with Reclaim the Streets protesters that private cars should be banned from city centres. The poll is the first to find an outright majority – 57 against 38 per cent – in favour of allowing only buses and taxis to enter city centres. The poll comes after a long hot spell which has pushed traffic-generated ozone pollution above World Health Organization danger levels.'

Such reports signalled that transport issues were now unfolding in a very different climate of public opinion than had prevailed in the 1980s.

Figure 23 Andrew Lees speaks for Oxleas Wood, 1994
© Steven Bridge

Transport alternatives in the form of better public mass transit systems, or wider provision of dedicated bike routes in town and country, were finding positive and growing support among all sections of the population. That twentieth-century icon of freedom and modernity, the private car, was speedily losing its gloss.

Because many of the stumbling-blocks to a radical shift in attitudes to transport were essentially cultural, or to do with manipulating images and emotions rather than facts, FoE turned to countering the 'hidden persuaders' in advertising or PR. Advertisers expertly persisted in glossing cars as designer objects of desire, charmed passports to personal freedom, warm expressions of individuality and sexual potential. It was hard to connect this image with the mobile pollution factory, the deadly weapon, the land-hog, the gridlocked battery cage, all the downsides that showed up if you set the car in a different light.

Pious rhetoric and statistical litanies had little chance of supplanting the advertised image and message. Nor could groups like FoE afford to fight fire with fire by buying media space and competing on level terms. It had to be a case of boxing cleverer. In effect, this initially meant playing the visual techniques of advertisers back at the media with a changed message. An eye-catching Saatchi and Saatchi promotion for British Airways, which used hundreds of human figures to bit-map the world, gave birth to a succession of anti-road actions which borrowed the same device.

It was logical in many ways that FoE should take this more visual turn for, more than any other issue, transport campaigning engaged with 'media theatre' forms of ethical advocacy that played to the camera and microphone. From imitations of advertising art, this visual strand began to evolve into projects that courted artists who could create completely unexpected and original art scenarios with added ethical value and point.

The organisation had had close connections with eminent artists in many media before, but in a very different role. When he had taken over as Director in 1984, Jonathon Porritt had invited Victoria Cliff-Hodges, a seasoned music administrator at Covent Garden opera who was keenly interested in nature and environment, to join the organisation. Her assignment was to build a capacity in FoE to set up public events and exhibitions with an arts and entertainment spin that would draw new kinds of recruits to the green cause and shed its dowdy image. He also saw revenue-earning potential in such events.

Cliff-Hodges began by trying to organise major music concerts, building on FoE's 1970s track record as a pet cause of celebrity musicians. But she hit a snag. Most of the stars who were interested in collaborating had concert schedules tied up years in advance. The days of the impromptu good-cause gig were by now pretty well numbered, though (as Bob Geldof and Band Aid were to prove a little later on) the genre still had aces up its sleeve.

The Arts for The Earth – or TATE – set out on a different tack, with an auction of cartoons donated by top cartoonists including Charles Schulz, Ralph Steadman and Gerald Scarfe. Nick Bonham of the prestigious Bonham's auction house sponsored and staged the event. *Times* cartoonist Barry Fantoni helped set it up. The 120 lots fetched over £20,000. One guest, the Chairman of the Heinemann publishing house, suggested to Cliff-Hodges that she have the cartoons photographed before they were taken away, so they could be merchandised in book form, too. Heinemann subseqently co-published the book, *Earth Mirth*.

From then on there was a Bonham's auction every year for several years on varying themes: a book illustration show, paintings and other art on tree themes and a rainforest art exhibition. The latter show coincided with a spectacular Rainforest Ball in the Hippodrome theatre in London's West

End in 1987. The celebrity guest list included Mick Jagger, Elton John and David Bowie.

Ballgoers wore fancy dress appropriate to the theme, from banana skins and palm trees to fake gorilla skins. Leading West End restaurants served themed dinners to the guests beforehand and singers and comics staged impromptu cabaret performances. The event raised over £150,000. Though publicised and packaged in slick and glamorous style, such events delivered hard-hitting punchlines once the punters had been lured in.

Film and theatre events followed, including a specially created one-off performance of John Osborne's play *Look Back In Anger*, starring Dame Judi Dench, Emma Thompson and Kenneth Branagh. The production was a sell-out at ticket prices of up to £1,000 each. Channel 4 screened the event, further multiplying its audience and spreading the environmental messages that formed its prelude. The arts made excellent sense, it appeared, as a channel for outreach, publicity and fund-raising. They also brought new allies into the picture like Olivia Harrison, wife of former Beatle George. Her special interest in water and wetland issues had been spotted and encouraged by Andrew Lees. The Harrisons subsequently co-hosted several film events in London which netted big revenues.

Arts events of varying size and scope were staged by TATE outside London, too, in venues up and down the country. There was a major production at the Stratford Memorial Theatre, a concert event at the Aldeburgh Festival venue of The Maltings, a recitation by Dame Peggy Ashcroft in Salisbury Cathedral and a host of small-scale events each of which raised funds and communicated a green message.

Victoria Cliff-Hodges moved on in 1990 to start the Green Screen environmental film festival and archive, and TATE was disbanded not long afterwards. A fruitful connection with the arts was soon resumed, however, albeit in a different context. This time artists and their work were seen as potential agents of environmental care and advocacy in their own right, rather than as box-office draws or charitable come-ons.

A demonstration of the proactive use of art took conspicuous shape in the Grey Man of Ditchling, an anti-roadbuilding protest in Sussex in early July 1994. This was when the Tour de France bike rally, the world's largest spectator sport event, was due to pass through the South Downs at Ditchling Beacon, near Brighton, on the penultimate leg of a UK detour. The media circus the event towed in its wake was gargantuan. FoE Brighton teamed up with local anti-roadbuilding citizens' groups to take advantage of the cavalcade by protesting plans for a South Coast Route road scheme from Folkestone in Kent to Honiton in Devon.

Through the local School of Art, these groups sounded out Brighton-based cartoonist Steve Bell and land artist Simon English. The duo were asked to create a figure on a hill site visible from the Tour route at Ditchling. Intended to parody the ancient Long Man hill figure at nearby Wilmington, it was installed in a single morning by volunteers from FoE

Figure 24 The Grey Man of Ditchling, August 1994
© David Townend/FoE

in Brighton and London, local anti-road groups in the South Coast Against Roadbuilding coalition and Brighton art school students.

Resplendent in Y-fronts and complete with traffic-cone headdress, the 150-foot chalk figure of John Major greeted the Tour as it crested the Beacon. The largest political cartoon ever published, it also grabbed the attention of the media, especially (and oddly) the Tory loyalist press. Along with its 'NO MORE MAJOR ROADS' caption it got splash coverage in the *Times*, *Daily Telegraph* and *Daily Mail*. The Grey Man also appeared in several prime-time TV slots, not just in Britain but in Europe and round the world, too. Some two years after it had appeared and disappeared (the work was carefully designed to leave no impact on the land), it was still cropping up on TV screens as far afield as South Africa and Brazil.

This staying power testifies to the effectiveness of art as a campaign medium. The cost of the commission was less than £2,000, yet the notional cost of buying the media space the Grey Man gained would have run into six figures. But did the Grey Man play a real part in influencing the subsequent decision to scrap controversial sections of the route?

Such connections are hard to measure, but it did seem that here was a manner of campaigning that resurrected the spirit of invention and devilry of the Schweppes bottle dump. Its visual impact allowed it to creep under the defences the media had put up against being used as a billboard for the messages of pressure groups. It helped that it was funny but, equally important, it was visually arresting and freshly conceived, not so much a publicity stunt as a public art spectacle, with an additional payload as a work of environmental advocacy.

Other ways of recruiting art, native wit and creativity to green ends have been devised by former FoE principals Angela King and Sue Clifford. But their group, Common Ground, has more participatory forms of creative synergy in mind. It promotes art projects that reinforce bonds between local people and their environment. For instance, Common Ground's Parish Maps project prompts rural communities to diagram their familiar surroundings in such forms as tapestry or mural, emphasising the things that matter to them. They work either by themselves or in harness with a visiting artist or craftsperson.

One nationwide Common Ground project is dedicated to preserving orchards that harbour rare or unusual fruit varieties. Another encourages people in towns and cities to trace the origins of local street names to underlying natural or agricultural features that urban development supplanted, so re-establishing a bond between the land and the communities and individuals who live and rely on it.

A shift in social and communal values in favour of environmental stewardship has always been high on the list of FoE's goals and ambitions. Building a sense among people of being stakeholders in what happens to their surroundings, rather than passive bystanders or exploitative consumers, has been the main angle FoE has taken on another heated 1990s issue, which has overlapped with transport issues to a very significant extent.

This is the fate of Britain's SSSIs and the natural biodiversity they harbour. In 1989 a Royal Society for Nature Conservation report reviewed habitat destruction in the UK and in 1991 Wildlife Link issued *SSSIs – a health check*. In 1994 an alliance of conservation groups produced *Biodiversity Challenge*, an attempt to influence an official Biodiversity Action Plan then shaping within the Department of the Environment. In the same year, FoE published a survey of threats to SSSIs in England and Wales and a report (*Gaining Interest*) demanding new laws to shore up protection. All the signals from these studies were alarm signals.

But what could be done? Everyone already knew that wildlife habitats in the UK were still being lost, and it was clear that campaigning in a general way was unlikely to work. A more specific approach was needed; one that would get the issues into the media, show that there was a problem and provoke a political response.

Tony Juniper, by now heading the biodiversity and habitats campaign, thought that focusing on a handful of threatened sites might be the answer. In April 1994, The Magnificent Seven campaign was born. Building on the roads and peatlands campaigns, and using one or two long-running campaign issues such as the proposed Cardiff Bay Barrage, FoE sought to mobilise its own community groups and those of other organisations in an effort to defend some of the country's finest remaining wildlife habitats. The plan was again to expose in a hardhitting way why the Wildlife and Countryside Act was so weak on SSSI protection and then to use the public's interest in, and better understanding of, the issue as a level for new legislation.

The campaign quickly gathered momentum. By October, a joint legal challenge against the British government by FoE and the Ulster-based Friends of Ballynahone Bog led to the repeal of a peat-digging permission due to the government's failure to comply with the European Habitats Directive. The bog was saved. Soon after, Somerset FoE activists, led by Richard Dixon, succeeded in turfing Lord Hanson's Amey Roadstone Corporation out of an ancient woodland site in the Mendips earmarked for quarrying.

The campaign for new laws attracted many followers. Sir William Wilkinson, Derek Ratcliffe and Norman Moore, the old guard of Britain's conservationists, joined FoE in delivering a damning report on the extent of the damage and the threat to key nature resources across the country to the Department of the Environment and organised a press conference in the House of Commons. At the other end of the spectrum, the direct action campaigners also joined the fray. In early summer 1995 they moved onto the Selar Farm SSSI in West Glamorgan, which plans for an open-cast coalmine threatened with oblivion. They also began to show a healthy interest in another focus of that campaign – the Newbury bypass.

Meanwhile the interest and activism whipped up by the campaigns on the seven sites was helping to build support for proposed new legislation to beef up the discredited 1981 Wildlife and Countryside Act. By late 1995, 260 MPs had signed a parliamentary petition supporting an FoE-drafted Wildlife Bill, and in December that year the bill saw its first reading in the House of Commons after being picked up by James Couchman MP.

Yet though many established conservation groups evidently see this issue as their own, they differ widely among themselves in the interpretations and solutions they offer. As Bill Adams notes in *Future Nature*: 'In the UK the message about local communities and conservation has been accepted by some parts of conservation, but not by others.' He traces a traditional divide between landscape conservation for natural beauty, recreation and the like, and nature conservation for safeguarding habitats or biodiversity.

This divide, says Adams, also separates attitudes to people and how their stake and say in conservation is regarded. He draws a parallel with the dilemmas faced by wildlife conservationists in Africa, who have learned to recognise that conservation must contribute to solving the problems of the rural poor who live day to day with wild animals, or those animals will disappear. FoE's involvement in the SSSI debate continues to hinge emphatically on a people-oriented attitude to conservation.

Jonathon Porritt sees this emphasis as a distinguishing feature of all FoE's campaigns:

> There's no doubt, that Friends of the Earth has always sought to justify its interventions in the political process in terms of how it will work out for the human species. The important thing has always been: if you can't make it work for human beings, it won't work. You don't have to justify it in balance-sheet terms or some of the other more short-term, narrow and self-interested measures of human benefit people use today. You can use longer-term, more enlightened yardsticks. But you've still got to put that into the equation.

As FoE continued to take stock of its position on all these issues and more, a familiar pattern of debate began to emerge. In the apparent stand-off between those who urged a higher-profile, more nature-centred and more confrontational agenda and those who championed a high-risk but arguably less glamorous quest for people-centred sustainability, weren't there shades of the Druids versus Wise Users stand-off that had set David Brower on the FoE trail in the first place?

Brower initially had no global issues like climate change or ozone holes to consider, no acid rain, no rainforest destruction. Or rather, if these things were in the air they didn't enter the public reckoning. Brower clearly recognised, however, that the only prize ultimately worth striving for, whether the root motivation was nature-centred or people-centred, was the blue-green nugget pictured in the Apollo images from space, and its frail-looking envelope of life. So long as we befriended the Earth, did it matter so much which side of our split collective nature discharged the endlessly demanding and rewarding duties of friendship?

16

Friends and relations

> You who are conscious and individual should never do to the chaotic and universal anything you wouldn't want it to do to you.
>
> – Boris Pasternak, *The Childhood of Lyuvers*, 1916

Following David Gee's departure in January 1992, the search began for a successor at the top. Gee left of his own accord, exasperated by in-fighting over campaign priorities. Iris Webb, who was shortly to end her long service as a Board member then a Trustee, could sympathise. She felt that such conflict had too often sapped the group's energies at the very times it most needed its wits about it.

A common bone of contention in internal set-tos was a feeling that Friends of the Earth had to shed its cerebral corridors-of-power image and challenge a perception 'out there' that FoE came a poor second to Greenpeace as an environmental publicist, mover and shaker, and now needed to adopt a higher public profile. It was true that, though Greenpeace was also feeling the pinch of the recession, its growth outstripped the rest during the boom of the 1980s. The French Secret Service gave Greenpeace an unscheduled boost in 1985 by blowing up the group's flagship the *Rainbow Warrior* in Auckland harbour, where it was preparing to sail into France's nuclear test zone. Crewman Fernando Pereira perished with it.

In the months immediately following this act, Greenpeace's UK membership almost doubled. Yet Greenpeace was no longer the *enfant terrible* of direct action it had been of old. It had gone through much the same process of professional development as FoE, including determined moves to upgrade its science credibility and its effectiveness as a formal lobbying force.

The non-violent 'Blokes in Boats' guerilla tactics that brought the group fame remained its trademark but David McTaggart and other principals now saw direct action as a weapon of last resort, to be used only when science and sweet reason failed. In Greenpeace there were inevitably some who felt this more reticent line was a mistake, that it threatened to erode the moral high ground and media centre stage it had staked out.

Figure 25 Charles Secrett at the 1996 Newbury Rally
© Nick Cobbing/FoE

Similarly there were factions in FoE that saw more direct and headline-grabbing campaign styles, rather than subtle new angles on sustainable development, as the way to go. Somewhat against the grain of this tendency, Charles Secrett emerged from a formidable shortlist to take over the helm in January 1993. His definition of FoE's agenda for the 1990s was sustainable development pursued in a national or local context, taking the Earth Summit agreements at face value as a mandate for radical change at the grassroots.

Secrett had the confidence of the Board and Trust as a safe pair of hands to cope with managing the various factions at the centre. He also knew the organisation back to front and had distinguished himself as a campaigner. What he was now offering, however, was no easy option. 'I felt we weren't thinking up new strategic approaches, not evolving, adapting as we had,' he says. 'We needed to see light at the end of the tunnel. I made it clear what my vision for the organisation was when the Board interviewed me, and I think it rang a bell for them, for the local groups and most of the staff.'

In practical terms, though, the first step was to get the organisation's finances in order. In common with other membership-based voluntary groups, the recession had hit hard. Being without a director for so long – Dave Gee had left fifteen months earlier – hadn't helped matters either. Cutting out non-essential expenditure was a start. But redundancies were inevitable. Twenty-seven posts, nearly a quarter of the total, had to go.

This sort of move can destroy a campaigning group, where passionate, talented people work incredibly hard and long hours because they believe in what they do. Working for FoE is as much a way of life as a job. But, in a quite remarkable way, options were discussed positively and openly, and the difficult choices made. Given the bitter experience of the early 1980s, the whole process of saying final thank-yous and good-byes to so many colleagues and friends was a tremendous testament to everyone involved, and a mark of how much FoE had grown.

At the same time, the opportunity was there to set FoE on its new path. Restructuring – that dreaded corporate word – allowed the organisation to reduce its administration to the essential minimum. Campaigns were reorganised, and campaign, research and information posts were protected from cuts as much as possible. A new sustainable development research team was established, to pioneer FoE's work on cross-cutting sustainability themes, and demonstrate the economic and social benefits of the environmentalists' agenda across Whitehall, and to politicians, industry and the public alike. Plans were made to create a Local Campaigns Department, sending the strongest possible signal to the local groups and the outside world that local campaigning was just as important as campaigning at national or international levels – and would be resourced properly.

Alongside these developments, another change began. This one was a change of attitude and tactical approach. With many others, Secrett believed strongly that FoE had to stop being so London-centric. Underwood Street staff should think and act as if local group volunteers and activists were colleagues to be respected and supported, and more fully involved in helping shape the group's strategy and campaign plans. Building up FoE's strength across the country became a top priority.

Doing so gave the organisation the opportunity systematically to develop campaign tactics such as parliamentary campaigning and citizen action across the range of its work, and not sporadically on one or two issues. Over the next two years, FoE campaigned vigorously to help the Home Energy Conservation Bill (an initiative of the Green Party and the Association for the Conservation of Energy) become law. Considerable efforts were made to reduce the ridiculously high VAT rate on energy conservation materials and energy saving goods and services; and an unprecedented alliance formed with poverty, housing and community groups to ensure that high VAT rates on energy use did not penalise the disadvantaged. The group drafted its own Wildlife Bill, to bolster the hopeless habitat protection sections of the 1981 Wildlife and Countryside Act and drafted the Road Traffic Reduction Bill with the Green Party and Plaid Cymru. Both bills quickly attracted extensive all-party political support in local authorities and in Parliament.

The combination of local action with national campaigning and media work paid off in other ways. Campaigners were encouraged to involve supporters and members of the public, as well as local group members, and to develop campaigns that gave people the opportunity to become environmentally active citizens – as voters, consumers, shareholders, investors, employees and employers, and, as FoE had traditionally emphasised, in their own life-styles.

The supporters' magazine, *Earth Matters*, became a crucial campaign and information tool. Redesigned and beefed up considerably, with fewer articles on FoE specifically and more on environmental topics written by outside writers, the emphasis is on stimulating supporters to think about what is going on – and act. A 4-page 'Take Action' section sends a very clear message – there is a lot that you can do as individuals working together to help the world become a better place.

Secrett says,

> One of the great beauties of living and working in a capitalist democracy is that people have the power to change what industry and government do. Often they don't believe it. But if no one buys a product, who is going to make it? If politicians believe a measure is popular, they will support it because they want to get elected. One of our jobs as campaigners is to give people the confidence and the opportunities to make a difference. And that means getting the right information to the right people in the right way at the right time.

Secrett's plans had the additional advantage of defining clear and appropriate differences between FoE's style of campaigning and that of peer organisations, while setting a direction-finding lead away from the becalming effect of the Earth Summit aftermath. Could such a lead be what the environmental movement as a whole is waiting for?

Pete Wilkinson wonders whether there ever was such a thing as an environmental movement. 'I suspect at times it's just a bunch of organisations out there vying for the same bit of charitable spending,' he says with an almost straight face. But if there is, he adds, it has to change with the times.

> The environmental groups have to be inventive, imaginative, out there all the time at the cutting edge. Sometimes I think the situation out there is flat, everyone is bored stiff, nothing's happening. If anything's happening I think it's in industry or business more than anywhere. Just talk to anybody in the movement that's been around for a while and they're saying, where's the next initiative coming from, where's the next big quantum leap?

Jonathon Porritt sees the idea of a movement that develops in waves, each rendering the last irrelevant, as a misreading of history.

> In the old days it always looked that way but it just hasn't happened. The conventional conservationist route to looking after the environment is still there. Friends of the Earth and Greenpeace didn't supplant it, they merely came in and offered another way of doing it. Look at RSPB, CPRE, even the National Trust. They're probably a great deal stronger that they ever

> were before Friends of the Earth or Greenpeace arrived. Groups like Friends of the Earth opened up new areas of lobbying for them that they'd have never had otherwise.

Nor does Porritt see the emergence of radical single-issue groups in recent years as a threat to the status quo.

> It now seems clear, that you need an array of different organisations using different tactics. Young people have to discover all over again the currency of anger and indignation, fear, hope and vitality, as if it were freshly minted. It's preposterous to assume that the legacy that us lot or the generation before might pass on would be a quick route through to the level of understanding that got us where we are. That's not how human learning patterns work.

Porritt feels that the increasing sophistication of the movement now allows for a much clearer sense of each group's positioning within it and greater understanding on the part of those who subscribe to green ideas about what each group can offer. He is sure FoE and peer groups like WWF or Greenpeace are here to stay.

> I'm not being complacent about the future. Friends of the Earth always faces the challenge of assessing its own usefulness, its role in society at any one time. So its chiefs can't sniff complacently and say we'll be here forever, we can do things our way. And they're not doing that. They're going through the greatest shake-up in their history.

FoE's position in relation to other environmental groups is unlikely, however, to change much in general terms. Despite any confusion the public may sometimes feel about the differences, people who know the movement well are clear enough about where the key distinctions lie. Chris Rose, a veteran of all the 'Big Three' groups, is forthright on what he sees as the main distinction between FoE and Greenpeace. 'Greenpeace gives images, Friends of the Earth gives information,' is how he sums it up. 'Greenpeace sees things in black-and-white, Friends of the Earth says, well, there's a lot of shades of grey in here. But alas, by the time they've embarked on the first sentence of explanation, most of the audience has dropped off to sleep!'

Charles Secrett sees no point in contesting this reading but rejects its conclusion.

> Greenpeace *is* more an organ of action, and Friends of the Earth is more driven by ideas. But Greenpeace works on a narrower range of issues, dominated by marine issues. We work across the widest possible range of issues. We don't see how you can deal with any single issue in isolation from the others. Time and again we led the way in shaping and developing the agenda of the green movement as a whole, in terms of introducing new issues to campaign on that others then see the relevance of and follow suit.

Secrett is committed to an agenda of development and social justice issues in the Rich World–Poor World context, an area that Greenpeace has almost always shied away from, even during the Earth Summit circus. 'We've largely tried to avoid it,' says Rose, 'we don't want to get trapped inside that. The last thing we want to do is get drawn into a position where there's great rafts of agendas all smothered by huge amounts of discussion.'

This view is arguably the antithesis of that of WWF, which (except in the USA) has changed its full name from World Wildlife Fund to World Wide Fund for Nature. Despite this nominal change of emphasis it remains at heart a charity wedded to wildlife and habitat conservation, with a strong accent on education. Charles Secrett sees the WWF core agenda as:

> wild species and habitat conservation in a fairly traditional manner. Though WWF also works on resource management and environmental protection issues, it's not a top priority. In tactical terms, WWF works closely with industry and as a consequence is not particularly good at changing it. But WWF also works closely with governments and government agencies on practical 'wise use' projects. It has always achieved good things this way.

Jonathon Porritt, who also serves as a WWF Trustee, holds a different view. He regards WWF's caseload of humanitarian sustainable development projects tied in with educational initiatives in the North and South as assets which deserve more credit inside the green movement than they commonly get. 'As an organisation I think WWF has done a very brave job in seeking to put across conservation themes in a humanitarian light, often in ways that infuriate the animal rights or protectionist lobbies. Their views and approaches will in future cause more major confrontations.'

A fly in this ointment from the point of view of detractors is WWF's close association with commerce and industry and its blithe confidence in the good intentions, or 'enlightened self-interest', of the market sector. WWF's view is that industry will not, in the end, kill the goose that laid the golden eggs by stripping natural assets to the point where income, jobs and profits can no longer be had from them. Yet killing the goose is exactly what international trade allows capital investors to do with impunity, for once one location has been exploited to destruction the profits can simply be banked and re-invested in another spot ripe for development.

Greenpeace's position on animal welfare issues is harder to discern than WWF's. In the USA, Greenpeace almost pulled itself apart in the 1970s over the question of whether or not it was permissible for the Native American Inuit people to continue to hunt bowhead whales for a living in the Arctic. Since then, coincidentally or not, its agenda has apparently steered clear of such issues, though much of its early fame had been based on campaigning against seal culls and the like.

Turning to the differences between the Big Three groups in terms of their membership profile, it is apparent that WWF members have a much younger average age than the other Big Three groups. 'That's because they've concentrated on education and done brilliant work in it,' says Iris Webb. She sees the membership of respective organisations as distinctive identifying features that can be summed up in style terms.

> Greenpeace is very cool and trendy. It has a strong fashion thing, like it's cool to be in Greenpeace. Friends of the Earth tends to be seen as more straight and worthy, maybe more brainy. But on the other hand the poor old groupies don't get much of a look-in in the Greenpeace setup, they're not asked to be activists, just to give money.

Colin Hines of Greenpeace concurs:

> I see it as a bit like the difference between the Beatles and the Rolling Stones. You know, we tend to do offshore stuff, they do terrestrial stuff, they've got a network of groups working within the classic lobbying and action mode, we've tended to just use supporters as a fund-raising resource, we haven't had that group structure. But we've also had a very highly organised international structure, whereas theirs is looser. We work more internationally, partly because of the boat thing, and we don't take anybody's cash except as membership dues.

Hines is reluctant to make value judgements. 'For me it's a spectrum, you need action on a whole range of levels, I've never believed our way has to be the best way, I just happen personally to enjoy the way this organisation does it.' '*Vive la différence!*' appears to be a view Hines shares with the grassroots, for many FoE local group members are also subscribing members of Greenpeace or WWF, not to mention RSPB and the rest of the old-guard organisations.

In view of this overlap, are the style differences Hines and Webb describe an illusion? Why shouldn't the groups combine forces more? After all, all three work to protect the selfsame natural world. In fact, FoE and Greenpeace quite frequently collaborate on particular fronts, notably in recent years over nuclear issues, while campaigners from both groups sit together on joint bodies like Wildlife Link.

Charles Secrett sees the plurality as a plus. 'We share some objectives with the other groups but we work in completely different ways on different issues we see as priorities,' he points out. Even so, a blurring of the distinctions between the different organisations can lead to 'brand' confusion.

Jonathon Porritt is impatient of generalisations about the overlapping territorial claims of one group over another:

> If you actually did a timesheet of hours spent and cash spent on different activities, you'd see clear differentiation between the resources Friends of the Earth puts into certain kinds of activities and those the others put into theirs. The only thing that really gives you any real indication of how an organisation does its work is resource allocations. Specifically, if you look at the vast whack of money WWF puts into education, the navy of boats Greenpeace supports, the huge investment in time and money that Friends

> of the Earth puts into local group activities, you'd see at once that very different routes have been taken.

Where confusion remains, much of it comes from perceptions of the green movement in the media. A love-in between the mass media and environmental groups during the late 1970s and 1980s was a crucial factor in establishing a popular power base for FoE, Greenpeace and the rest. The media also grew to become powerful pro-environment persuaders in their own right, especially the TV documentary genre. The 1984 film report *Seeds of Despair*, for instance, alerted the world to famine in Ethiopia against the background of a civil war fought largely as a proxy Cold War contest.

The humanitarian response to the programme's harrowing images of environmental and humanitarian disaster was a revelation in people power. It was orchestrated around pop star Bob Geldof's heroic Band Aid initiative, and saw showbiz entertainers taking on a hitherto unimaginable role as global providers of social and environmental care. But from this high spot and another brief ratings blip during the Earth Summit, media attention to environmental concerns has declined in the 1990s.

Above all, TV programme controllers have shown less and less interest in documentaries on green topics. And apart from the work of a handful of broadsheet correspondents, the Geoff Leans, John Vidals, Nick Schoons and Fred Pearces of the business, there has been a dearth of press stories based on a knowledgeable view of the issues. Yet readership surveys show that informative environmental stories are still in steady demand. Jonathon Porritt has strong views on this.

> It is a source of anxiety to all of us in the green movement that we're so badly served by those who have such influence. We've suffered an enormous amount by people who simply don't understand what is happening to the green idea. They can't track it, they don't really know what to do with it. They don't know whether to treat it as a special vested interest, or as a broad social movement or as a kind of confrontational voice in society or whether to treat it as mainstream. And they're not interested enough to find out or work it out.
>
> But you know, in a way it doesn't really matter. When I was at Friends of the Earth or in the Green Party, we used to tear our hair, distraught at the difficulties we were having with the media. Whereas the other processes, the ways people arrive at decisions in their own right, are going on far more actively than the media ever reflects. The changes proceed almost despite the media.

More distributed forms of media technology like the Internet, or the capacity even the smallest environmental groups now have to desktop-publish newspapers on a personal computer have also, in Porritt's view, made a difference to the media equation. 'I think they help a lot. I think we'll see nothing but improvements in that area. Look at the way the anti-roads movement, for instance, has used information technology.'

Chris Rose agrees that straight news and other media coverage of environmental stories has declined. He feels, however, that in a certain sense the environment story never really went away. 'It's not just the specialised environmental programming that counts,' he remarks. 'It comes out all over the place, in recreational programming, social affairs coverage and so on. And everywhere you go, environmental thinking has become normalised in all forms of communication.'

Rose also thinks that people are taking more environmentally conscious actions in their lifestyles even without necessarily thinking of them that way, whether that means installing low-energy lightbulbs or buying ozone-friendly fridges or organic vegetables.

> The marketing men, who look at long-term changes in society and business, realise that. They take it as read that the consumers of tomorrow are greener than yesterday's. I'm not saying I feel good about every aspect of it. But I don't go along with those who see this dreadful media blackout closing in around them. Some parts of it will come back and hopefully it will be better.

Tom Burke strongly rejects any idea that an ethical cause can expect to own the media agenda:

> The spotlight is always looking for somebody who's dancing. Whether you attract the spotlight your way matters, you have to manage that system. But it's very two-edged. Those who live by the spotlight die by it. As time goes by your sense of what is and isn't achievement changes. It has to be tilted towards getting lots of headlines but that doesn't last. That's what it's about to work with the media. It wants to move on, it doesn't want to look back.

The criticism that green organisations have become enslaved by fossilised expectations on the part of supporters and the media has lately been voiced by a growing band of sceptical observers on both sides of the Atlantic, though with different conclusions in tow. The 'contrarian' view of several British commentators, including economist Wilfred Beckerman and former environmental purist Richard North, is that commerce and industry are not the Evil Empires pressure groups have made them out to be in the past. They have skeletons rattling in their closets but they alone have the spending power to solve many of the problems our industrial past has visited on us.

American journalist Mark Dowie presents a different point of view in his recent polemic *Losing Ground*. He blames the incursion of marketing specialists onto the ethical pressure-group scene and in particular the introduction of direct-mail and telephone fund-raising appeals into the equation, for an increasing lack of bite and relevance he perceives in the green movement of today. He believes that the more individualistic movements for social rights and justice, such as the feminist, gay rights or indigenous peoples' movements, hold the high ground because they don't fudge the issues.

Amid these mostly gloomy prognoses and the reduction in charitable or ethical giving attributed to the economic recession, forward planning within FoE is focused on steering a distinctive course away from received roles and stereotypes. Charles Secrett believes the future for the green movement is bright if it becomes more fully involved in what he calls 'the politics of society'. He says,

> The biggest challenge to the status quo will come from the environment movement. The shift is inevitable, because humans lose out when they trash nature. We should push for integration of environmental thinking with public and private sector economies sooner rather than later because the sooner we get it, the less pollution, the less habitat destruction, the less waste of valuable resources and the fewer extinctions there will be. Environmental problems cost vast amounts of money and make our lives worse. Prevent them, and human welfare soars.

A controversial part of the Secrett formula for FoE's future development is to face up to the political flip-side of 'greening' society and the shape of things to come. According to Secrett,

> There's room for debate and discussion about exactly how one goes about stimulating such processes, but I don't think the need for them can be contested. The fight will be about politics, because it will be about who has the power to decide what's quality of life or what isn't.
>
> We're in the business of trying to drive *solutions*. Of course we're up against powerful vested interests, both political and corporate. You see this particularly in countries like Indonesia or Brazil where it's almost impossible to see where political rule ends and commercial interests begin. But it happens everywhere. So when we talk about sustainable solutions, we're not just talking about what should happen in the South but about what happens here too.

As Secrett was making these observations, news was coming in from Nigeria about the imprisonment and judicial murder of writer and activist Ken Saro-Wiwa, in retribution for a protest he had organised against ecological disasters created by oil extraction in the homeland of his own Ogoni people. Ken Saro-Wiwa is not the only green advocate who has been victimised by oppressive regimes in the global South during recent decades. Others are Sunderlal Bahaguna in India, Kenya's Wangari Maathai, Marcos Chan Rodriguez in Mexico City and another local hero who paid with his life, Chico Mendes in Brazil.

The achievements and sacrifices of these outstanding men and women, the Browers, Carsons and McTaggarts of their nations and generations, are emblems of a new wave of pro-environment activism arising in the South, much as it arose in the North twenty-five years ago in the form of groups like FoE. Charles Secrett observes,

> So often, we hear politicians from the South say: this sustainable development sounds fine, but why aren't you doing it, too? That's what stonewalled the proposed Forests Convention at Rio and came close to derailing the

> Biodiversity and Climate Change Conventions. But it's a fair question, even when it comes from the mouths of politicians in the South who are no great fans of democracy. It's evident that the South can't change unless the North changes, too.

Secrett's observations relate to a growing general awareness that while a climate doomsday is a threat whose origin and antidote reside largely in the global North, another doomsday scenario driven by poverty and population overload is also coming rapidly to a head down South. All key international bodies now recognise parallels (and some perceive all manner of direct links) between steep population growth in the South and continuing industrial and economic growth in the North. Forced to export a lion's share of domestic product to repay foreign debts, developing country economies cannot guarantee needy people in poor neighbourhoods adequate land or livelihood. Lacking economic buffers, poor families simply cannot survive if their numbers are small.

As Hari Sharan commented in a *New Scientist* article of 1994: 'Of course the growth of Southern populations at current rates creates personal, social and ecological problems. If this growth continues it will certainly lead to global problems sooner or later.' Yet missionary attempts to urge family planning miss the point. Sharan explains,

> It would be downright suicidal to limit one's family to two children, when gathering fuel for cooking may need between two to six hours a day; when bringing water for humans and animals may require several treks of several kilometres a day; when every hand available must struggle on a small plot of land for meagre returns; when the only source of support in old age is the sons and when the chance of survival of small children and babies is still low because of lack of clean drinking water.

Add to this recipe for human and environmental overload the much bigger environmental burden caused by the profligate lifestyle of the North, and few gamblers in their right mind would take bets against serious crises of instability in nature, society and economies during the century to come.

Whether the crisis is 'South-led' or 'North-led', or a bit of both, its effects will be global. One ominous scenario is that disadvantaged people from the South might sweep across the major boundaries with the North, such as the Mediterranean or the Panama isthmus, to claim their fair share of resources by direct action. The EU takes this threat seriously and has earmarked billions of dollars of development investment to create a buffer-zone of relative affluence in North Africa.

Another demographic grey area is future assimilation of the newly democratised states of Eastern Europe and the former Soviet Union into the European whole. An urgent need before this can happen is to raise environmental protection standards in industry. The environmental quality owed to the citizen also has to be raised closer to Western levels if another environmental refugee situation is to be avoided. Average life expectancy in Moscow fell by ten years in the decade leading up to 1989, due mainly to lack of decent healthcare, poor diet and rampant industrial pollution. Only a radical overhaul of industrial infrastructure in the East can fend off further decline in the future. And such change presupposes radical political and economic transformation.

FoE's new agenda means intervening in the political process as never before. Says Charles Secrett: 'I think genuine public participation as a fundamental part of democracy is the truly radical message of the Earth Summit.' He points out that nobody need look further for an example than Principle One of the Rio Declaration, which states: 'Human beings are at the centre of concerns for sustainable development. They are entitled to a healthy and productive life, in harmony with Nature.'

FoE is evidently determined to act on this mandate. What they and many others argue from its basis will directly challenge political structures and the powers-that-be in Britain and in the world. Though devoid of poetry, the Earth Summit Agreements are as revolutionary in their way as anything that Adam Smith or Karl Marx ever wrote.

17

Promising the Earth

> What are global problems but an accumulation of local problems? Safe water, clean air, safe land, real forests, a quality of life where you can look forward to waking up in the morning. So many of the things I see you are fighting for in Britain, we are also fighting for at home. The same company you are fighting against goes to Ireland, goes to Malaysia and goes on to Africa.
>
> – Chee Yoke Ling, FoE Malaysia,
> John Preedy Memorial Lecture, September 1992

Even as sights were being set on future goals, the key issues in the organisation's existing portfolio were still alive and kicking, and demanding constant attention. The 'nuclear graveyard' planned by NIREX for Sellafield was still on the drawing-board. Friends of the Earth contested the company's planning applications to build a 'rock laboratory' with vigour. The 'laboratory', it maintained, was a Trojan Horse: once so much financial investment had been made, NIREX would not pull out, even if Sellafield were not found to be the best site.

Once again, FoE campaigners fought a nuclear inquiry, in partnership with the Cumbrian local groups. Once again, its witnesses – including NIREX 'supporters' – performed superbly, their evidence unrebutted. The company was forced to acknowledge the quality and detail of FoE's case and to present additional evidence. NIREX simply failed to respond to the most important points in FoE's evidence; even a cursory study of their replies gave a clear impression of who won the argument. FoE totally demolished NIREX's case.

As the campaigners await the Inspector's report, due in October 1996, they have a bigger victory to celebrate. In 1995, the Government had announced that it would make no further investment of public money in the nuclear industry beyond privatisation, and refused to subsidise the construction of new nuclear reactors. It meant the end, says Pad Green, of the nuclear dream. Issues like the management of radioactive wastes would run for as long as those wastes remained active. But the final withdrawal of official support from the nuclear industry was a plain and simple vindication of FoE's long-haul opposition to nuclear power development. Green remarks,

> If in 1977 we'd said this would happen by 1995, people would have laughed themselves silly. But that's what we have achieved. We've proven that nuclear power is unnecessary. We've always proposed alternative ways to supply energy and now there's a real renewable energy industry. The campaign focus for the future is on making the renewables industry grow quickly, and making sure that nuclear is not needed to fight climate change.

Pad Green holds the view that the British nuclear industry would do well to climb down from its pedestal and admit that safer waste storage technologies, rather than risky 'disposal' options based on highly uncertain premises, are the only sensible way forward in the short run. It might even corner lucrative European and world markets by capitalising on existing strengths in the area of radioactive waste storage.

This proposition is an example of FoE's preparedness to work with industry, government, or any other interest group in a constructive mode, rather than seek to combat or roadblock them. But such relations can only thrive within an agreed framework of environmental imperatives that all parties recognise and heed. As part of the new agenda of political relevance it is now bent on pursuing, FoE is ready to argue for just such a formal contract between citizen and state on environmental quality and security – clean air, pure water, uncontaminated land, unadulterated food – as fundamental rights guaranteed by the state for one and all. It will also spell out what it believes the balance of power should be between national government and local authorities when it comes to decisions which are apt to affect the environment. 'We must never forget that it's the poor who usually live downhill, downwind and downstream. Without such guarantees, the enjoyment of a clean and healthy environment will become increasingly the preserve of the wealthy who can afford to escape the messes we create,' argues Charles Secrett.

Since such a formula is impossible to deliver under the UK's present unwritten constitution, FoE has begun to campaign for a Bill of Environmental Rights. It has helped found a new coalition conceived by Jonathon Porritt and Richard Sandbrook, among others. Known as Real World, the coalition also includes WWF, Oxfam, Christian Aid, Charter 88 and various other social justice and church groups.

Jonathon Porritt says Real World's mission is to answer the question: what does sustainable development really mean? '*Not* sustainability as environmentalism redefined or development as regards only the Third World,' he explains. 'We want to know what *sustainable development in the UK* means. How do we integrate social justice, democratic renewal and responsible development and give them political leverage, something which isn't happening here now?'

Part of the answer must lie in the electoral process. In common with Real World and growing numbers of MPs from all the main parties, FoE has begun to urge a national referendum on proportional representation as a way to step up pressure for these changes. It regards the simple majority electoral system as a licence for the major political parties to ignore public concerns over environmental problems. Charles Secrett points out,

> No matter how high these concerns may mount, there's little parliamentary pressure on the party in power to improve environmental policy because smaller parties with strong pro-environment manifestos never make the breakthrough. So we see lack of proportional representation as a critical block to achieving a sustainable environment within which a prosperous and just civil society can thrive.

Sorting out political concerns has now become, according to Jonathon Porritt, a precondition of FoE being able to deliver on ecological sustainability across the board. 'I daresay Friends of the Earth will always remain at heart an organisation dedicated to finding a better pattern of living for the human species on Planet Earth,' he insists.

Charles Secrett acknowledges the risk of becoming tarred with political brushes but insists this need not happen.

> We're not a political party, we're a public interest pressure group. But to those who say you shouldn't be political, I'd reply that's nonsense, it bears no relevance to what a public interest pressure group does. Probably the only non-political human beings I can think of are monks or nuns who meditate in some cloister and have no contact with society. We have to be political because all the things we want and all the things we don't like are determined by the politicians. They set the rules by which society plays.
>
> What we'll never be, however, is party-political. Friends of the Earth can never allow itself to be accused of getting into bed with any particular sectarian or ideologically determined political viewpoint, nor of turning its back on any political party. There may be some so repugnant that we never want to have anything to do with them, such as the National Front. But if we stay true to our independence and our first principles, we'll be okay.

Secrett believes the current agenda of FoE and the way it is going can be summed up in a maxim. It is that there are no environmental problems and solutions (except over geological timescales), only economic, social and political problems and solutions. He explains,

> Environmentalism is about looking after conditions for *all* life, not just wildlife but also people. One of the reasons why the environmental agenda got trapped in a ghetto was because whenever it came to major decisions on the Budget or the ballot box or when investment decisions were made, environment suddenly got shoved onto the back burner.
>
> Despite a high level of popular concern over green issues, they were still seen as fringe concerns that could be ignored at will, issues somehow

> divorced from the business of living in a modern-day economy. What our new agenda is demonstrating is that being an environmentalist is about showing how, if we look after the planet, we meet our own needs and still leave room for the fulfilment of our aspirations.

Another side of the FoE's agenda is an emphatically positive outlook. 'We're *for* things, we favour our type of progress, we believe the world can be better.' Says Secrett, 'Too often environmentalists have been caricatured as being only interested in the worst sort of gloom and doom prophecies. People who are good at articulating what's bad but not so good at presenting a vision of what life could be like at the positive pole. That's most unfair when it's applied to FoE,' he argues.

Secrett feels that today's agenda can build on the group's achievements by continuing to come up with real and practical proposals, not just for solving environmental problems but also for helping to cope with social and economic issues like unemployment, healthcare, the relationship between the citizen and the state, knots that a conventional way of doing things has repeatedly failed to unravel.

> To take just one example: the environmental agenda could deliver hundreds of thousands of jobs within ten to fifteen years if pursued in such sectors as energy, agriculture, transport, pollution control and use of materials. We know this from comparable experiences in Germany, Sweden, Holland, Japan and the USA. Every person who escapes the dole saves the exchequer £9,000 a year. No sound politician can ignore that.

Approval for the Secrett formula within the FoE family has not been unanimous. Charles Secrett acknowledges this. 'Campaigners sometimes find it hard to see why we should go down this political path,' he concedes.

Some indication that FoE fully intends to maintain and raise its profile as a newsworthy pressure group can be gleaned from the recent appointment of former Greenpeace campaigner Uta Bellion to the key post of Campaigns Director. She says that what she finds most interesting about working for FoE is the opportunity it offers to work through local networks.

'That's much more difficult to handle,' she admits, 'but I see it as an essential way for people to link up together to deal with the issues that matter to them, rather than say to them this a big international topic and you've got to work on Issue X and nothing else.' Bellion also feels it will be important for FoE to strengthen its capacity to network issues internationally, too, in the years to come.

Less contentious than the political items in Secrett's Tomorrow File are new ideas for strengthening connections between the centre and the local groups. Secrett subscribes to the idea that campaigning at local level is just as important as campaigning at national level, no more and no less. Richard Sandbrook strongly believes this emphasis is right:

> Even though it creates all sorts of tensions between central and local interests, the unique strength of Friends of the Earth is in its network. The big

> challenge for environmental groups today is to solve problems locally which then have national and international consequences. There isn't a short cut because if you'll excuse the cliché, all environmental problems are local. Frameworks need to be there nationally, regionally, internationally. But there's this tendency to try and solve everything with all these ridiculous Conventions nobody takes any notice of because they've grown from the top down, not from the bottom up.

Secrett shares Sandbrook's view and wants to see it reflected in the organisation's strategies, tactics and operating procedures. 'We've set a limit on growth at the centre,' he explains. 'We only want to get so big at Underwood Street and no more. We'll look for future expansion out in the regions, in focal points around the country.' How will FoE further this decentralising process without diluting its identity or sapping its unity?

Part of the answer lies in the heavy investment the organisation has made in a new digital information system, known as FOEnet, which networks Underwood Street with the regional offices and disseminates information to the wider public over the Internet. FOEnet grew out of Andrew Lees's vision and forms part of his legacy to the organisation. An unabashed computer addict, Lees believed that accurate, telling information, often presented as a map, was a key distinguishing feature of FoE's output.

It was Lees who initially persuaded the organisation to make what has now been some ten years' worth of cumulative investment in information technology, including databases, geographical information systems and a presence on the Net. The first and second stages of FOEnet's development – the installation of the hardware and software at Underwood Street and the linking up of all the regional offices to FOEnet – is now complete. The third and final stage is to extend FoEnet facilities to the local groups, giving them access to the centre's extensive information resources, email and the Internet. Then there is the potential to broaden access still further, with outlets in public libraries and schools. Charles Secrett calls FOEnet 'one of the most exciting roads we're going down in empowering people, helping them take control of their lives and find out what's around them.'

FOEnet's programmers recently created an interactive mapping interface for the Government's database of industrial pollution, which has set new standards for access to public record information. By typing in a postcode, anyone can obtain a customised map, centred on their own home, showing the local factories that are monitored by the Government. Clicking on one of these factories brings up lists of all the recorded emissions and permitted pollution levels for that site. In the first week the site went online, more than 59,500 users logged on and 3,476 interrogated the database. In the whole of the previous year, just 760 had people visited the regional Pollution Inspectorate offices to seek out the same information in printed format.

The potential of the Internet to marshal support and keep pace with developments in a fast-moving campaign has recently been demonstrated

Figure 26 FoE's watchful presence ensures that the violent scenes of Twyford do not recur at Newbury © Andrew Testa/FoE

with the campaign to stop construction of the A34 Newbury bypass. Daily updates about the situation on the ground generate huge interest, and of course there is an interactive map of the site, giving information on the legislation, ecological assets and other relevant issues on this key campaign.

The Highways Agency announced on the eve of Parliament's annual recess in August 1995 that the bypass scheme would go ahead, cutting through nine miles of superb English countryside, including ancient woodlands, watermeadows, heath, bog and downland. The decision flew in the face of new Government guidance which criticised existing assessment methods for failing to account for new traffic generated by roadbuilding schemes themselves. The then Transport Minister Brian Mawhinney had announced a review of all road schemes to take into account induced traffic and the extra congestion it was likely to cause. FoE found that the Government's own figures showed that traffic through Newbury would be back up to pre-bypass levels within five years of the scheme's completion.

Land clearance contractors for the Newbury scheme began to tool up in late 1995, ringed around by huge contingents of private security guards. By January 1996, clearance had been chronically delayed thanks to highly organised passive resistance by protesters along the route. In February, FoE released new figures showing that the bypass would cut a mere two minutes on average off journeys along the existing urban route,

Figure 27 Newbury protester manhandled by security guards, 1996
© Nick Cobbing

a trifling gain compared to the environmental damage the scheme would cause. The Highways Agency claimed that the figures (again based on their own official statistics) were 'misleading because they were taken out of context'.

This cat-and-mouse game of claim and counter-claim was matched by thrust and counter-thrust on the field of what had now come to be styled 'the Third Battle of Newbury', referring to two English Civil War battlegrounds located along the bypass route. The protesters had a subplot, a European Directive that prevents tree felling during the songbird nesting season. Delaying tactics were therefore the order of the day.

The war of nerves even began to affect the men in hard hats. In January, several members of the contractors' security gangs defected to the side of the protesters. The rest were conscious that roughneck tactics of the kind used at Twyford were unlikely to go unnoticed at Newbury. For FoE has found a useful role it can play without flouting its 'within the law' mandate. It has marshalled a continuous watch on events by impartial witnesses trained to record incidents or confrontations for potential review as courtroom testimony. The presence of the legal observers has meant that violence to individuals has been kept to a minimum, and that evictions of protesters from their camps has largely been carried out within the law.

Meanwhile the blanket coverage generated by the campaigners resisting the clearance work on the SSSIs and other wildlife habitats was beaming

Figure 28 Anti-Newbury bypass protester, Penn Wood, 1996
© Alex Macnaughton

the continuing destruction of the UK's countryside into almost every home in the country. Tony Juniper says that the communications opportunity was unprecedented:

> It was the best chance we've had in recent times to show people what is still happening to our countryside. If we are going to have a sustainable economy, our society will have to make some difficult choices. Those choices were there for everyone to see at Newbury, and the campaign will undoubtedly prove a turning point because of that.

The dilemmas a modern society faces in meeting people's need for mobility while conserving its biodiversity were sharpened by FoE's solutions-led campaign at Newbury. Detailed research by the independent Metropolitan Transport Research Unit mapped out how alternatives to the bypass might solve local traffic problems at less cost to the Exchequer and without damaging the important wildlife and archeological sites that the road was going to ruin. These results proved critical in moving public opinion against the road and put those arguing for the scheme on the defensive. Indeed, shortly after the clearance work started, the local Liberal Democrat MP and transport ministers abandoned technical arguments for the road in favour of a smear campaign against the protesters.

Materials collected by FoE legal observers was catalogued at FoE's HQ in preparation for the hundreds of court cases that would follow the Newbury protests. Under Elaine Gilligan's coordination, FoE's aim of putting environmental rights on the political agenda was forcefully put into practice through a campaign for people's right to peaceful protest in defence of nature.

Relations between FoE and single-issue groups engaged in transport issues had by this time taken a healthier turn. Roger Higman says,

> We had a period of conflict and controversy after Twyford. There was a lot of slanging on both sides. But they've changed and we've changed. There was an element of jealousy – not just in Friends of the Earth – that had to work its way through. We've all realised that's pretty stupid and now there's a sort of *rapprochement*, we know what we can do for one another.

Despite this upturn in relations with other groups, Roger Higman eyes the ongoing showdown over Newbury warily, as well he might. 'It's a battle,' he says, 'but I don't think it will be a Twyford. It's dangerous going into a fight supposing that it's going to be the same as it was last time. We know we're not going to make the same mistakes, but we may not get the same opportunities either.'

As these pages go to press, transport and land use issues continue to dominate the Act Locally agenda of FoE in the towns and countryside of Britain. But the Think Globally side of its mission made a surprise appearance on the back doorstep from the early weeks of 1996 onwards, with the news that the British Isles are well in line for loser status in the lottery of global warming and climate change.

Doomsday Rides Again seems to have been a recurring theme in the history of FoE, for time after time, just when environmental affairs seem to be relatively lacking in drama, external events have conspired to thrust old-time environmental nemesis back into the headlines. According to forecasts published by the Intergovernmental Panel on Climate Change (IPCC), in 1994 the world had between forty and fifty years to stabilise greenhouse gas emissions before it ran 'quite high risks of irreversible climate change'. But with each succeeding forecast this time-frame has contracted.

In Britain, a summer of extreme drought in 1995 led to water rationing in most regions, then gave way to a severe and long-drawn-out winter. Was climate change due to global warming already a reality? In December 1995, the IPCC (which represents the work of 2,000 top meteorologists round the world) concluded that the balance of evidence suggested that human activities are contributing to an increase in global temperatures.

The consensus among scientists had previously been that the British Isles would be one of the few places on the planet to escape global warming, thanks to the mixing effect of a deep-sea current of very cold ocean water in the Atlantic between Greenland and Scotland. This, it was thought, would absorb any local warming caused by the greenhouse effect. Now this comforting prospect has been marred by much bleaker news.

In 1996, for an unprecedented third year running, a massive tongue of ice known as the Odden Feature failed to materialise off Greenland. Some oceanographers believe that in normal years the melting of the Feature causes huge amounts of fresh water to be sucked down into the deeps, feeding the submarine current. That deep salt current acts as a natural pump, drawing the Gulf Stream along its regular course. The Gulf Stream brushes the British Isles and moderates land and air temperatures as it passes.

In an *Independent on Sunday* report, Geoffrey Lean quoted the view of Dr Peter Wadhams of the Scott Polar Institute in Cambridge (who coordinates a European Commission research programme on the Odden Feature) that global warming is to blame for the Feature's disappearance. He says consequent weakening of the Gulf Stream is likely to make UK winters progressively colder, even as the world heats up. It seems we have by no means heard the last of climate change hereabouts.

On all the issues, from nuclear power to climate change, or from transport to nature reserves, on which FoE has made a principled stand in recent years, the policy current has sooner or later shifted its way. Sometimes this connection is not so visible to the public or media gaze as it is to the organisation's member-activists and campaign professionals. To maintain a popular support base and buffer its finances beyond fear of setbacks, FoE always needs to make its role as an agent of progressive environmental change and care evident to a widening public.

But those who follow environmental issues closely will be aware that FoE's position has almost always been vindicated in the long run. This success is a just reward for the organisation's dedicated achievers in every arena of activity, from parish pump to Parliament. Their satisfaction reflects in FoE's regular maintenance of membership and subscription support at a time when most other ethical pressure groups are losing ground.

How would Charles Secrett respond if one of the newer green pressure groups now emerging in the East or South asked him: what are the key lessons we can learn from FoE's experiences since 1971? Short of promising the Earth, in other words, what can an independent environmental group reasonably expect to promise and how should it go about it? He replies:

> Never forget who you are. That's the key thing. Never be beguiled by other people's view of what your message is or what environmental campaigning's about. Because embedded in our outlook is a conviction that things can be achieved in lots of different ways. If you want to emulate us at our best, be an open, transparent, participative group. Combine intellectual rigour with passionate commitment: don't get the facts wrong.
>
> Create structures founded in devolved decision-making and don't fall into the trap of letting a few people decide what everyone else should do. Take a holistic view. Acknowledge that the environmental agenda is a complex one and that the problems aren't going to be resolved easily. Be patient and have the humility to recognise that you haven't got all the answers, that you're not always going to be able to do things all on your own. So listen to others and work with them.

Finally, be independent-minded. Never be co-opted by a sectarian or partisan point of view, no matter how superficially attractive it may seem under the conditions or circumstances of the day. I'd say those are the essentials.

Appendix

25 years of campaigning success

Note: This list is inevitably selective. Some years are sparse because of poor records. This compilation is indicative. We have striven for accuracy, although errors may have crept in.

1971 FoE publishes three pioneering books, *The Environmental Handbook*, *The Population Bomb* and *Concorde: The Case Against Supersonic Transport*. Five national country groups attend the first FoE International meeting in Paris.

1972 National Packaging Day is marked by FoE local group demonstrations nationwide. FoE's lobbying helps set the agenda at the Stockholm UN Conference on the Human Environment. The 'Save the Whale' campaign kicks off with a demonstration outside the International Whaling Commission (IWC), the publication of *The Whale Manual* and a protest rally in Trafalgar Square. Two hundred MPs sign an FoE early day motion calling for an import ban on whale products, and the Pet Food Association announces a voluntary ban on using whale meat. FoE opposes mining giant RTZ's plans to mine copper in Snowdonia National Park, through lobbying and legal work. FoE presents Prime Minister Ted Heath with a bicycle after traffic jams made him late for work.

1973 The Government announces plans to build 32 pressurised water reactors (PWRs), and FoE announces its opposition. FoE publishes *World Energy Strategies* and *Packaging in Britain*, showing the huge potential for recycling. FoE drafts the Endangered Species Bill and submits 150 amendments to the Government's Control of Pollution Bill. There are now 73 local FoE groups.

1974 FoE's Endangered Species Bill gains a second reading in the House of Lords, despite Government opposition. Local groups build fifty giant paper mountains outside council offices for the 'Great Paper Chase' campaign and work with Age Concern to insulate pensioners' lofts in more than 50 towns. FoE opposes Torness nuclear plans. The Whale Products Boycott campaign begins: a giant inflatable whale is sunk outside the IWC, attracting global press coverage. FoE publishes *Wheels Within Wheels*, the first exposé of the powerful roads lobby, *Losing Ground*, on agriculture and self-sufficiency, and *Britain and the World Food Crisis*. The campaign to protect allotments nationwide is launched.

1975 FoE and the RSNC draft the Wild Creatures and Wild Plants Bill. FoE groups, in cooperation with local authorities, create hundreds of allotments on disused sites around the country. 3,000 activists march past Downing Street on FoE's Bike Day. FoE submits detailed evidence to the Royal Commission on Environmental Pollution (RCEP) inquiry into nuclear energy, the House of Commons Committee inquiry into the motor industry and the M16 Inquiry – and creates a public outcry over Windscale expansion plans. Sperm oil demonstrations mark FoE's Whale Week. Staff total 18; there are 140 local groups; and 12 countries attend the FoE International meeting in London.

1976 Minister Tony Benn chairs a televised public meeting between British Nuclear Fuels (BNFL) and FoE. FoE publishes *Getting Nowhere Fast*, combining a devastating critique of current transport policies with promoting solutions. Local groups begin setting up recycling schemes. After intense campaigning by both FoE and BNFL, the Government backs FoE and announces a public inquiry into Windscale expansion plans. Supporters number 5,000.

1977 FoE joins with other groups to campaign against the Canadian seal hunt. An FoE conference in Salzburg reveals that Israel has hijacked 200 tons of uranium. FoE leads objectors at the Windscale inquiry on cost, proliferation and environmental risk, detailing the case against the THORP reprocessing plant; it also holds anti-whaling demonstrations outside Japanese and USSR embassies. FoE publishes *Paper Chain* and *The Politics of Urban Transport Planning* (with Earth Resources Research) and *Soft Energy Paths* (with Penguin).

1978 FoE persuades the Government to reduce official traffic forecasts by 50 per cent, and to decide against buying PWRs. The Goverment launches a National Anti-Waste Programme in response to an FoE-initiated study on returnable containers. FoE submits evidence to the Foster Committee on Heavy Lorries. 12,000 people say 'No! to Windscale' in an FoE Trafalgar Square rally. FoE publishes the *Bicycle Planning Handbook* and campaign manuals on cycling and pollution. FoE co-sponsors the first International Green Film Festival in London. FoE takes out the first private prosecution under the Endangered Species Act and submits evidence to the M25 public inquiry.

1979 100 FoE local groups stage energy conservation events. FoE publishes *Rethinking Electricity, Torness: Keep it Green, Economic Growth* (on allotments) and *Frozen Fire* (on the dangers of liquefied natural gas). 6,000 cyclists are at the 'Reclaim the Roads' rally and block Whitehall for several hours; 12,000 people demonstrate against whaling. FoE has 16,000 supporters.

1980 FoE cyclists block British Rail offices in protest at BR's decision to ban bikes from commuter trains. 20,000 people protest at FoE's anti-nuclear rally in Trafalgar Square. FoE has 150 local groups and a budget of £250,000.

1981 FoE launches 'Paradise Lost', a campaign to protect British wildlife habitats. FoE dumps half a tonne of bones at the Department of the

Environment to protest at the trade in endangered species. The House of Lords Committee on Science and Technology agrees with FoE's evidence and concludes that a nuclear expansion programme is not needed. FoE and BANC publish *Cash or Crisis?*, forecasting the failure, and huge expense, of the Wildlife and Countryside Act. 15,000 people attend the Last Whale Rally in Hyde Park, the largest wildlife rally ever. 50,000 supporters 'Send a Tin to No. 10' to back recycling, and an FoE-drafted Returnable Bottle Bill is introduced in the Lords.

1982 Vale of Evesham FoE saves Donyatt Cutting from becoming a rubbish dump. 10,000 people join an FoE/IFAW anti-seal-hunt rally in Trafalgar Square. The CEGB admits that reprocessing AGR fuel is a mistake. Bournemouth and Poole FoE groups begin an ultimately successful campaign against oil drilling in Poole Harbour.

1983 FoE publishes hidden Government evidence of massive damage to wildlife habitats since 1945 and drafts a Natural Heritage Bill to replace the Wildlife and Countryside Act. Led by Avon and Birmingham FoE, local groups run insulation and recycling schemes. FoE and IFAW organize the second anti-seal-hunt rally in Trafalgar Square. FoE stages sold-out Green Rallies in London, Bristol and Leeds, featuring Ralph Nader.

1984 FoE launches the first UK campaign against acid rain; it also publishes *An Investigation into the Use and Effects of Pesticides in the UK*. FoE discovers that DDT is illegally on sale, and stocks are withdrawn. FoE publicises damage to 133 Sites of Special Scientific Interest (SSSIs). FoE leads objectors on safety issues at the inquiry to build a nuclear reactor at Sizewell. The publication of *The Gravedigger's Dilemma* triggers a wide-ranging debate on nuclear waste disposal. The Commons Environment Committee accepts FoE's evidence that the UK should cut acid emissions by 30 per cent. Lobbying by the Freedom of Information Campaign and FoE secures amendments to the Food and Environment Act so that pesticides test data is disclosed.

1985 FoE publicises damage and destruction to 245 SSSIs in evidence that persuades the Commons Environment Committee that wildlife law needs strengthening. FoE publishes *Motorway Madness*, revealing threats to 100 SSSIs. FoE launches the first international tropical rainforest campaign, focusing on trade and aid issues, with consumer and taxpayer actions; it publishes *Rainforests: Protecting the Planet's Richest Resource, Timber! An Investigation of the UK's Tropical Timber Trade* and *The Tropical Hardwood Product List*, detailing unsustainable timber products nationwide. FoE organises a mass rally to protect Oxleas Wood SSSI in south London. FoE's first 'Forest Alert!' nationwide tree surveys reveal widespread acid rain damage and prompt the Government to order an official national survey, which confirms FoE's findings. FoE gives evidence to a Lords Select Committee investigating dumping at sea, emphasising the importance of land-based alternatives which the Hazardous Wastes Inspectorate later endorses. FoE publishes *The First Incidents Report*, persuading the Health and Safety Executive to overhaul pesticide poisoning monitoring. Broadlands FoE reveals widespread and illegal mercury pollution of River Yare. *World in Action* focuses on the FoE and Cambridge FoE campaign to save Swavesey Meadows from drainage by Anglian Water; the campaign with the Wildlife Trust saves the most important site, Mare Fen. FoE brings two representatives from the Northern Marianas to the London Dumping Convention, floats a massive 'tropical island' on the Thames past the Convention and exposes Japanese plans to dump nuclear waste near the islands. FoE and FoE Scotland launch a campaign to protect Duich Moss SSSI from drainage on Islay. FoE hosts a major conference on 'Safe Routes to Schools' for local councils, and co-organises National Bike Week. There are 200 local FoE groups.

1986 FoE supports local campaigns against nuclear plant sites at Dungeness, Hinckley Point, Druridge Bay, Wylfa and Sizewell. The Commons Environment Committee's report supports much of FoE's evidence and is highly critical of nuclear waste management. FoE launches the Cities for People campaign. FoE publishes *Better Roads for a Better Economy* and *Colossus of Roads*, exposing the myths that building roads is a cure for unemployment. FoE's radiation expertise helps reassure thousands of anxious callers and informs media commentary after the Chernobyl disaster. FoE works with MPs to successfully harden Labour and SDP policies against nuclear power. FoE publishes *Energy Without End*, the first analysis of the potential for renewable energy in the UK, and *The Good Wood Guide*, which lists sustainably produced tropical timber products and acceptable temperate alternatives endorsed by The National Association of Retail Furnishers. The European Commission endorses FoE's code of conduct for the tropical timber trade. FoE publishes *Nitrates: Boon or Bane?*, and after FoE's formal complaint the European Commission begins proceedings against the UK for breaching drinking water law. FoE hosts a major international conference on radiation issues; the proceedings are published by Wiley & Sons as *Radiation and Health: The Biological Effects of Ionising Radiation*; 800 scientists world-wide back radiation exposure limit reductions. Local groups raise £4,500 to support Sarawak FoE's logging road blockades. Avon and Vale of Evesham FoE persuade Wessex Water to prosecute a Tenneco factory over illegal pesticide discharges into the Severn.

1987 FoE's lobbying strengthens Labour, Liberal and SDP transport manifestos. FoE publishes *Air Pollution from Diesel Engines*, showing that the health risks are greater than those posed by petrol. An FoE national environmental checklist campaign is launched for voters to judge all general election candidates. FoE submits evidence to the Commons Environment Committee detailing river pollution from farm sewage, nitrates and pesticides. FoE publishes *A Hardwood Story: Europe's Involvement in the Tropical Timber Trade* simultaneously in London, Amsterdam, Bonn, Rome and Paris. FoE and Survival International's campaign persuades the Overseas Development Agency (ODA) to rethink aid for Indonesia's Transmigration Programme. Over 100,000 people join an FoE/CND march on the anniversary of Chernobyl. FoE publishes secret evidence showing the Government ignored scientific advice on pesticides threats when lobbying against European control and misled Parliament in 1985. The Commons Agriculture Committee upholds FoE's evidence and condemns the Government for weak controls on pesticides. FoE's Radiation Monitoring Unit reveals the extent of Chernobyl fallout and helps farmers in Wales and the Lake District win compensation. It discovers extensive contamination of Lynn Trawsfynydd, and reveals that Trawsfynydd poses the greatest public health risk of nuclear stations in the UK. The National Radiological Protection Board accepts the case for reducing 'safe dose' levels for workers and the public. FoE works with MEPs across Europe to strengthen directives on power station pollutants, car exhausts and diesel emissions. Survival International and FoE work to stop aid funding for dams, iron ore and timber in Amazonia. Roads Minister Peter Bottomley launches FoE's 'Kids Alive' campaign. FoE gives evidence to the Commons Transport Committee's inquiry into the roads programme, arguing for severe cuts and alternatives. FoE reveals acid rain damage to Windsor Great Park, and hosts an international conference on acid rain with the Goethe Institute. FoE publishes *Cause for Concern*, on wildlife threats from acid rain, and *Chemical Trespass: Whose Turn Next?*, detailing 149 pesticide-spraying incidents. FoE's campaign forces the Forestry Commission to admit acid rain damage to conifers. FoE distributes hundreds of thousands of copies of *The Aerosol Connection*, listing all CFC-containing aerosols and CFC-free alternatives. The World Bank announces increased support for tropical forest conservation, and the Inter-American Development Bank withdraws funding for a road in western Amazonia. The pesticide tributyl tin is banned. FoE's budget is £1 million; staff number 34; there are over 150 local groups.

1988 FoE gives evidence to the Commons Environment Committee Inquiry on acid rain, ozone depletion and global warming; the report endorses many of FoE's recommendations, including an immediate 85 per cent cut in ozone-depleting chemicals. The Government later agrees. FoE exposes Shell's continued sale of the pesticide aldrin, prompting the Government to announce a ban on aldrin from 1992. FoE launches the UK's first global warming campaign. FoE publishes *Tackling the Freight Problem*. Kirklees Council agrees to work with FoE on the first council environmental audit across all sectors. The Government abandons the fast-breeder programme and Dounreay by 1994. The Commons Agriculture Committee report agrees with FoE's evidence that MAFF had been negligent in handling Chernobyl contamination. FoE publishes *The End of the Nuclear Dream*, the popular *Green Consumer Action Guide* and *The Green Gauntlet*, with WWF and Greenpeace, challenging the Government to take environmental action in all areas of policy to wide publicity. FoE organises an international conference in London on ozone depletion, attended by many of the world's leading scientists. FoE reveals widespread illegal contamination of drinking water supplies in *Pesticides in Drinking Water* and formally complains to the European Commission. FoE highlights the huge subsidies for company cars; the Chancellor begins reducing them in the Budget. When the Nigerian Government invites FoE to investigate an illegal toxic waste dump, FoE exposes the international toxic waste trade scandal and forces tighter controls on hazardous wastes in Europe.

1989 FoE publishes *Safe as Houses* and *Uses of CFCs in Buildings* to show how the construction industry can curb CFCs; the Royal Institute of British Architects and the building trade applaud both. FoE exposes the Government's plans to relax pollution control standards for over 1,000 sewage works in the run-up to privatisation, causing a public and parliamentary outcry. FoE gives evidence to the Hinkley Inquiry on occupational risk and radiation dose levels. FoE reveals that some potato varieties contain twice the levels of tecnazene accepted internationally. FoE publishes *Once Is Not Enough*, a critique of waste management policies and practices, and, with WWF and CPRE, *Blueprint for a Green Europe*, which is aimed at voters in the European Elections. FoE helps the Kayapo Indians stage a huge gathering at Altamira in Brazil, to protest against dam-building, receiving massive global media coverage. The Lords European Committee report on energy efficiency cites FoE evidence eighteen times. FoE gives a keynote address to an inter-governmental conference in Helsinki, where the UK and other governments pledge to phase out CFCs by 2000. FoE releases a major exposé of the Government's obstructive environmental track record over ten years, to widespread publicity and political embarrassment; this is later published as *How Green is Britain* by Century Hutchinson. The European Parliament passes a resolution, based on FoE's work, to control tropical timber imports. FoE and UK 2000 launch the Recycling City initiative in Sheffield (later extended to Cardiff, Dundee and Devon) to demonstrate the potential for recycling. Over a hundred UK companies are awarded the FoE Good Wood Guide Seal of Approval. FoE helps coordinate over 120 local anti-roads groups against the London Assessment Studies. FoE publishes a guide to objecting to the proposed lowering of controls on sewage discharges; over 12,000 people formally object. Most recommendations from the Commons Environment Committee inquiry support FoE's evidence on energy policy and global warming. FoE launches the newsletter *The Powerline* with Greenpeace and starts a lobbying campaign directed at the City's financial sector, warning of the liabilities of nuclear power. FoE reveals that one in five people breathe air that fails WHO safety limits. An FoE and EDF-led coalition forces the World Bank and UK banks to drop loans for the Altamira dams. The *Sunday Times*, in an article based on FoE's work, exposes the destructive activities of UK transnational companies in rainforest regions. FoE launches the influential *Environmental Charter for Local Government* and reveals the extent of solvent pollution in Midlands underground water supplies. FoE in association with the *Daily Telegraph* publishes 51 regional recycling directories, which meet with massive demand. FoE's budget is £2.9 million; staff number 75; there are 270 local groups and 200,000 supporters.

1990 In association with the *Observer* magazine, FoE exposes 1,300 toxic landfill sites that threaten drinking water supplies; 10,000 people request FoE's local maps. The Government responds by instructing councils to compile contaminated land registers, and the National Rivers Authority (NRA) expands its groundwater monitoring. FoE presents its pioneering research on *Deforestation Rates in Tropical Forests and Global Warming* to the Intergovernmental Panel on Climate Change (IPCC). FoE groups persuade Sainsbury's, Safeway and Tesco to stock recycled toilet paper, and national newspapers to invest in recycled paper. Scores of local authorities pledge to stop using tropical hardwoods, following publication of FoE's *Good Wood Manual*. FoE publicises the ozone-depleting potential of HCFCs, marketed by the chemical industry as safe. FoE reveals pesticide contamination of ketchup and instant mashed potato brands. FoE launches a campaign with other conservation groups to protect threatened peat bogs from the horticultural industry. Later the Department of Trade and Industry Advisory Group recommends restrictions on the peat trade and the use of waste-derived composts instead of peat; HRH Prince of Wales commits his lands to peat-free horticulture. FoE's monitoring reveals ground-level ozone pollution in breach of European legal standards, and FoE calls for tighter vehicle emission controls. FoE joins with three fuel-poverty groups to call for an energy efficiency programme to help low-income households. FoE stirs an international outcry at the EEC environment ministers' meeting by releasing a secret US Government memorandum identifying the UK and Canada as potential allies to help block moves to cut greenhouse gas emissions. FoE stages a major international conference, 'The Rainforest Harvest', opened by the Prince of Wales and addressed by Brazil's Secretary for the Environment and the leader of Brazil's rubber-tappers' movement. The Government amends the Environment Protection Bill, following FoE lobbying, to boost recycling by councils. The Lords European Committee supports FoE evidence calling for a code of conduct for European logging companies and limits to unsustainable tropical wood imports. FoE presents *Getting Out of the Greenhouse* to the IPCC, showing that energy conservation and renewable energy programmes most effectively tackle climate change, to wide acclaim. FoE highlights bathing water pollution and lists beaches that fail European standards. FoE's study *London's Green Spaces: What Are They Worth?* forces the Department of Transport to admit that open spaces are not costed in road appraisals. FoE works with five trade unions to reduce worker exposure levels. FoE visits Belarus, at the invitation of the Belarussian Academy of Sciences, and helps uncover evidence of increases in childhood thyroid disorders. FoE releases monitoring results on the carcinogenic petrol additive benzene to coincide with the National Motor Show and calls for controls. HMIP confirms FoE's discovery of unauthorised radioactive discharges on Rainham Marshes SSSI. 20,000 households sign an FoE pledge to cut energy use by 20 per cent. FoE reveals that 392 sewage treatment works fail to meet pollution permits. The organisation's budget is now £5.5 million; there are 270 local groups and 220,000 supporters.

1991 FoE and the Policy Studies Institute publish *Reviving the City*, on achieving sustainable living patterns. 34,000 supporters write to the ODA demanding environmental and social controls on ITTO forestry projects; damaging projects are dropped or substantially improved, and a new project review process is set up. A joint FoE and local community campaign stops the proposed M40 city centre link in South Birmingham. FoE reveals that drinking water supplies for 14 million people are illegally contaminated by pesticides. The European Commission begins proceedings against the Government, following FoE's complaint. FoE publishes *The Peat Alternatives Manual* for the horticultural and landscape professions. FoE reveals that the Government failed to enforce acid rain pollution standards in order to lower costs for privatised National Power and Powergen. The Government agrees to fit speed limiters on heavy goods vehicles, a partial success for the carbon dioxide control campaign. FoE publicises 68 contaminated sites in London used for the production of 'town gas', many unknown to British Gas or local councils. FoE reveals that 160 SSSIs are threatened by new roads. FoE gives evidence to a Commons Committee Inquiry into eco-labelling; the Committee's report endorses FoE's recommendation to amend the Trade Descriptions Act to prevent false claims. FoE publishes *Coming Clean II*, a survey of 300 companies showing that voluntary controls on CFCs do not work. British Gas admits to degrading Ecuadorean rainforests, and pledges to take corrective action and consult with indigenous peoples, and the World Bank announces it will stop funding logging projects in primary rainforests, following FoE's campaign. FoE publishes the influential *Recycling Officers' Handbook* for councils and a report demonstrating that technology can cut transport carbon dioxide emissions by 30 per cent by 2005. FoE exposes secret plans for the East Coast Motorway, which threatens 16 SSSIs and 17 greenfield sites; Cambridgeshire, Yorkshire and Norfolk County Councils withdraw their support in face of FoE and local opposition. FoE publishes *How to Be a Friend of the Earth*, a handbook for personal environmental action. The European Commission upholds FoE's formal complaints that the Hackney M11 link, the M3 through Twyford Down and the East London River Crossing through Oxleas Wood had not been properly assessed for environmental impact and strongly criticises the Government. FoE lodges a similar complaint against the Newbury Bypass. FoE surveys reveal that barely 3 per cent of CFCs in abandoned fridges are recovered and that fish in 62 English and Welsh rivers are contaminated by persistent poisons above safety levels. FoE publishes *Off the Treadmill*, a blueprint for farming and the countryside. Twenty councils formally adopt FoE's Environment Charter; dozens of others adopt sections. Over a hundred local groups target DIY chains and their tropical hardwood use. FoE persuades the ODA to shelve a road scheme through a rainforest national park in the Central African Republic.

1992 FoE publicises Government data to reveal that 19 million people in the UK breathe air that fails EC safety guidelines. Thousands of FoE and Greenpeace supporters persuade HMIP to hold a formal public consultation into the THORP reprocessing plant at Sellafield. Every MP is sent a copy of *British Nuclear Fools*, detailing the economic and environmental case against reprocessing; a public inquiry results. FoE joins protesters in peaceful civil disobedience to stop the M3 extension through Twyford Down, until forced off the site by a court injunction. FoE exposes an international scandal of forgery, timber smuggling and corruption behind European logging companies plundering Ghana's rainforests; FoE's investigation, at the request of the Ghanaian Government, becomes a hard-hitting TV documentary. FoE exposes the Government's resistance to tough targets for banning all CFCs and publishes guidelines for phasing out halons in fire-fighting equipment; ICI agrees to halt production of the most common halon by 1993. FoE exposes the lax regulatory system allowing companies to discharge 23.3 tonnes of toxic metals into the sewers every day and publishes *River Pollution: A Sleuth's Guide*, explaining how people can uncover water pollution and take action. FoE campaigners at the Rio Earth Summit provide constant analysis of negotiations for the global media. The Climate Change Convention includes equity and historic responsibility clauses, as campaigned for by FoE, although the Forest Convention proposals are blocked. FoE publishes *Don't Throw It All Away*, a guide to waste reduction and recycling and *Less Traffic, Better Towns*, a popular manual outlining best practice for urban transport systems and reductions in traffic. FoE launches its 'Mahogany is Murder' campaign to stop Brazilian mahogany imports unless sustainably harvested. FoE presents comprehensive evidence to the RCEP's inquiry into transport and the environment, and to the Standing Advisory Committee into Trunk Road Assessment inquiry on road-induced traffic. FoE publishes *Whose Hand on the Chainsaw?*, which presents damning evidence of the destructive impact of UK trade and aid policies on rainforests. FoE also publishes *Dangerous Liaisons: Western Involvement in the Nuclear Power Industry of Central and Eastern Europe*, part of a three-year programme involving FoE groups throughout eastern Europe, and releases *Poverty, Population and the Planet*, a policy document on population and consumption.

1993 FoE publishes *Water Pollution: Finding the Facts*, a local activists' guide to water pollution campaigning, and *Critical Loads and UK Air Pollution Policy*, which highlights acid rain damage on ecosystems and presses for tougher power station emission controls. FoE targets National Power and Powergen, distributes accurate local damage maps to local groups in affected regions, and thousands of objections flood to HMIP. FoE submits a detailed report to the Government (*Sellafield: The Contaminated Legacy*), highlighting flaws in radiation testing and management on discharges and updates its original evidence against THORP for the 1977 Windscale Inquiry, demonstrating that its economic and safety objections hold true. FoE formally complains to the European Commission that opening the plant contravenes European law and prepares a detailed submission for the Government's official review of the prospects for nuclear power, calling for a coherent energy strategy based on energy efficiency and renewable supplies. FoE and FoE Brazil expose British timber companies trading in illegally felled mahogany. FoE co-hosts the first conference on urban traffic reduction with the Association of Metropolitan Authorities and follows up with regional seminars. FoE invites Amazonian peoples' leaders to the UK to lobby MPs and to support nation-wide demonstrations outside mahogany retailers by 100 local groups. *Forests Foregone: The EC Trade in Tropical Timbers and the Destruction of the Rainforests* is launched in Brussels and London. *Compost! A Guide for Local Authorities* demonstrates the potential for organic waste management. HMIP announces tougher acid rain emission cuts for National Power and Powergen by 2001 and calls for long-term clean-up plans. FoE's 'Stand Up for the Countryside' campaign has local groups nationwide acting as countryside watchdogs against damaging developments. People Against the River Crossing, Greenwich and Lewisham FoE and FoE publish *Change Here!*, detailing sustainable transport solutions to local congestion to protect Oxleas Wood SSSI: all local MPs endorse the study and oppose the road. FoE co-stages a rally at Oxleas Wood to protest against road threats; thousands of people form a huge map of the UK marking threatened SSSI sites. FoE organises a major international conference, *Delivering the Right to Know*, with US and European experts, on Freedom of Information environmental policy. Following FoE revelations, the NRA prosecutes chemical company Hickson & Welch for polluting the river Aire. FoE releases *Scab Wars: The Impacts of Organophosphate Sheep Dips on Farmers, Livestock and the Environment*, an authoritative assessment presented to MAFF. FoE gives evidence to the Commons Environment Committee inquiry into energy efficiency in buildings; their report cites FoE evidence 23 times and endorses many recommendations. FoE Cymru gives evidence to the Welsh Affairs Committee into wind energy: the final report supports FoE arguments for expansion. FoE accuses OFWAT of exploiting public concern over rocketing water bills to allow water companies to underinvest in pollution control (a charge officially confirmed in 1996). FoE local group campaigns stop damaging road schemes in Hereford, West Yorkshire (M62–M1 link), Preston (southern and western by-passes), Greater Manchester (M56–M62 link), south-west London (various) and around the M25. FoE focuses on the Government's failure to ensure councils provide contaminated land registers and publishes the popular *Buyer Beware*, a guide for prospective home owners to discover whether their home is built on contaminated land and which pollutants may be present. FoE releases a secret Government report revealing their aim of lowering water quality standards during EC negotiations. FoE launches *Conservation in Europe: Will Britain Make the Grade?*, an expert study revealing that 700 potential Special Areas of Conservation are threatened with damaging development, and is supported by five of Britain's most respected conservationists. FoE launches its Community Billboard opposite the Departments of the Environment and Transport, with the first poster reading 'ABOUT TO BULLDOZE 190 WILDLIFE RESERVES? THE NATION'S WATCHING YOU, JOHN'. FoE delivers a 150,000-

signature petition, 'Write off Nuclear Power', the largest ever, to the Government. FoE gives evidence to the Commons Environment Committee on recycling, reuse and waste management. FoE publishes the popular *Take the Heat off the Planet*, explaining how people can cut global warming emissions and fuel costs themselves and pressure retailers and the Government to act. FoE releases *From Dirty Man to Drittsekk: UK Acid Rain Policy*, detailing threats to over 1,000 SSSIs and a damning critique of control policies. FoE, RSPB, RSNC, WWF, Plantlife and the Butterfly Conservation Society launch a campaign to force the Government to set statutory targets and timetables for conserving threatened wildlife with the publication of *Biodiversity Challenge: An Agenda for Conservation Action in the UK*. FoE's response to the Government's consultation on air quality monitoring demonstrates system-wide deficiencies in coverage and accuracy. FoE ends the year by releasing a leaked ministerial letter, exposing the conflicts between the Departments of Environment and Transport over transport policies and habitat protection obligations.

1994 FoE launches 'The Climate Resolution' campaign through FoE local groups, targeting local authority areas for 30 per cent cuts in carbon dioxide emissions by 2005. The European Commission begins legal proceedings against the Government, following FoE's formal complaint, over failings in UK air pollution monitoring: the Government increases the number of monitoring stations and range of pollutants analysed. FoE uses ODA documents to demonstrate that the environmental impacts of the Pergau dam were not properly assessed before aid funds were given. Wiltshire FoE organises a major conference on sustainable transport solutions for the south-west region. FoE Crickhowell stops the A40 Bypass from damaging a SSSI and the Brecon Beacons National Park. FoE, RSPB and local campaigners win protection for Ballynahone Bog (from peat-cutting), Red Moss SSSI (from a rubbish dump), Canford Health SSSI (from a road), Asham Woods and Carmel Woods SSSIs (from quarrying) and the Severn Estuary (from a barrage). The threat of legal action from FoE and Friends of Ballynahone Bog persuades the Minister to save the Bog. Downpatrick FoE ensures Tree Preservation Orders are served on 174 acres of threatened woodland. FoE lobbying secures a review of nuclear waste management policy. FoE gives evidence to the Lords Committee inquiry into packaging and waste, arguing for waste minimisation, reuse and recycling. The Government issues PPG Note 13, supporting a long-held FoE position that the planning system should discourage car use. 10,000 supporters write to the Northern Ireland Environment Minister urging the designation of Ballynahone Bog and a halt to commercial peat extraction. FoE publishes *Planning for the Planet*, a popular guide on achieving environmentally sustainable development for council planners and officers. FoE gives evidence to the Lords Committee on Sustainable Development, outlining key public and private sector requirements and objectives. FoE publishes *Planning for Wind Power: Guidelines for Project Developers and Local Planners* to enable developments to incorporate strong environmental and social

obligations. FoE's air pollution monitoring patrol visits Cardiff, Manchester and Leeds to test health-threatening exhaust fume levels. FoE's acid rain cinema commercial, created by McCann Erikson, wins the BBC Design Award for graphic design. FoE, the Third Battle of Newbury and FoE Newbury hold a 1,000-strong protest rally at Donnington Castle, against the Newbury bypass. With Steve Bell, FoE creates a giant, 200 foot high 'Grey Man of Ditchling' caricature of John Major on a hill opposite the Tour de France, generating global coverage of the campaign against the south coast superhighway. FoE northern groups organise PRIDE, a week-long cycle protest tour of cross-Pennine communities threatened by road-building. FoE publishes *Compost II: A Guide for Farmers*, showing how to gain environmental and economic advantage from composting. FoE gives evidence to the Commons Committee inquiry into London transport air pollution in London: their report agrees that technology cannot cure pollution, only traffic reduction. FoE celebrates the IMF/World Bank 50th anniversary with the critique *50 Years is Enough*. FoE organises a 1,000-strong rally and mass Parliamentary lobby for the Home Energy Conservation Bill with the Association for the Conservation of Energy (ACE) and the Green Party. FoE publishes *Roads to Ruin*, showing roads threaten 39 SSSIs and 12 AONBs. Over 1,000 youngsters participate in FoE and the *Young Telegraph*'s Car Watch campaign. FoE Liverpool and the Vauxhall Health Forum prove health-threatening air pollution episodes from traffic in Vauxhall; the groups meet the Minister in London to press for traffic reduction, use the FoE Community Billboard and are invited to speak at a WHO conference in Madrid. FoE publishes *Losing Interest: A Survey of Threats to SSSIs*, detailing hundreds of sites, and launches FoE's Wildlife Bill in Parliament. The RCEP's transport report endorses many FoE criticisms of transport policy and procedures, as well as its recommendations for increased environmental protection and alternatives to new roads, the car and lorry. FoE publishes *Working Future? Jobs and the Environment*, detailing the employment potential in cutting pollution, conserving resources and environmental protection to wide acclaim by senior politicians. FoE and local groups carry out a peaceful demonstration in Heysham Port, targeting a ship with illegal mahogany imports. FoE co-hosts a Parliamentary press launch with ACE and Age Concern to promote the Home Energy Conservation Bill; many fuel-poor people are invited to tell their stories. Cumbria County Council rejects NIREX's planning application for a nuclear waste dump and forces a public inquiry. FoE submits *Beyond Safe and Economic* to the Nuclear Review, arguing for a comprehensive energy policy based primarily on efficiency and renewables; *Time to Face the Inevitable* argues that dry storage above ground is the only feasible option for nuclear waste, with the industry as a world leader in responsible storage and monitoring.

1995 The South Coast Against Roadbuilding coalition, led by local FoE groups, confirm their opposition to the superhighway proposed between Folkestone and Honiton by lighting Armada beacons for 230 miles along the coast. FoE and the *Independent on Sunday* magazine reveal the extent of health-threatening air pollution. The Commons Environment Committee inquiry on volatile organic compounds supports much of FoE's evidence and agrees emissions must be curbed to prevent health-threatening smog. FoE and FoE Scotland publish *An Environmental Assessment of Alkyl Phenol Ethoxylates and Alkyphenols*, the first thorough analysis of the threats posed by pervasive oestrogen-mimicking chemicals. WWF and leading environmentalists join FoE in pressing RTZ to stop a devastating mine in Madagascar. FoE and FoE local groups stage 'Mahogany is Murder' week, with campaign events and a dramatic cinema commercial. Cornwall FoE groups organise a major conference on sustainable transport solutions for the region. West London FoE challenges all aspects of the proposed Heathrow Terminal 5 at the inquiry. FoE gives evidence to the Commons Committee inquiry on waste fuels and cement kilns. Dozens of FoE shareholders appear at RTZ's annual meeting, forcing the board to admit the Madagascar mine's damage. FoE campaigners and activists 'recover' stolen mahogany products from retailers. FoE's Pollution Patrol reveals a major summer smog alert, to widespread coverage. FoE, the Green Party and Plaid Cymru launch their Road Traffic Reduction Bill in Parliament to set statutory targets for reducing traffic levels. FoE, the Third Battle of Newbury and local people stage a mass rally at Newbury. The European Commission upholds an FoE Cymru complaint that all oil and gas explorations require a full environmental impact assessment. FoE launches its environment and public health campaign, with *Prescription for Change: Health and the Environment*, a comprehensive analysis of public health threats from environmental degradation. British Energy cancels orders for the last two nuclear power stations. FoE pulls together an expert staff and independent witness team to be the lead objector at the NIREX dump inquiry at Sellafield. FoE Cumbria earn wide publicity for their county tour with a 14ft-high Trojan Horse. FoE releases secret data from English Nature, showing hundreds of SSSIs have been damaged between 1991 and 1994. FoE, with the support of WWF and the Third Battle of Newbury, release *End of the Road*, demonstrating that traffic management alternatives can solve Newbury's congestion. FoE releases a secret Government report showing that 42 of 80 UK wetlands protected under the Ramsar Convention have been degraded or are threatened. FoE launches the UK's first Chemical Release Inventory, using its geographical information system over the Internet, allowing people to discover which pollutants are produced, and where, by over 1,000 separate companies; HMIP then commits to do the same. FoE gives evidence to the Commons Committee inquiry into drinking water; their report supports FoE's position on pesticides and lead in water supplies. FoE's Wildlife Bill is introduced into Parliament. 32 councils back FoE's Climate Change Resolution. *GIS World* magazine names FoE's the 'Best Non-Profit GIS Site' on the Internet. FoE employs 110 staff, with 260 local groups and 180,000 supporters.

Index

Note: Figures in *italics* indicate illustrations.

Yes, I want to be a Friend of the Earth

Mr/Mrs/Miss/Ms____

Address ____

______________________________Postcode__________

RC96065713

Annual Membership

I enclose £16 ☐ Individual
£25 ☐ 2 people at same address/Family
£20 ☐ Overseas
£8 ☐ Youth/Student/OAP/Unwaged (delete as appropriate)

I'd like to make a donation of

£10 ☐ £15 ☐ £25 ☐ £35 ☐ £50 ☐ £100 ☐

I enclose £______ cheque/PO payable to *Friends of the Earth*
or please debit my Visa/Access/Mastercard

Card No ☐☐☐☐ ☐☐☐☐ ☐☐☐☐ ☐☐☐☐

Expiry date ☐☐

Signature ______________________________ Date__________

Regular Support

Your membership in the form of a monthly standing order will enable us to plan our campaigns even more effectively

Yes, I'll give a monthly gift of: £25 ☐ £20 ☐ £15 ☐ £10 ☐
£5 ☐ £3 ☐ Other £ ______

Name of my bank ____________________________________

Address of my bank __________________________________

______________________________Postcode__________

Current account number ☐☐☐☐☐☐☐☐ Sort code ☐☐☐

I would like my standing order to start in ☐☐ (month/year)

Signature ______________________________ Date__________

Bank instructions: Please pay the above amount on the 18th of each month to Friends of the Earth, Account No. 6500 6858, The Co-operative Bank p.l.c, 1 Balloon Street, Manchester M60 4EP. Sort code: 08 90 00 Please quote reference no:

Please send this form to: Friends of the Earth, Membership Department, FREEPOST, 56–58 Alma Street, Luton LU1 2YZ